The Complete Book of Fingermath

Edwin M. Lieberthal

Bernadette Lieberthal

Project Coordinator

McGraw-Hill Book Company

New York · St. Louis · San Francisco · Auckland · Bogotá · Düsseldorf ·
Johannesburg · London · Madrid · Mexico · Montreal · New Delhi · Panama ·
Paris · Sao Paulo · Singapore · Sydney · Tokyo · Toronto

IN MEMORY OF P.C.J.L.

Editor: Leo Gafney

Production by: Cobb/Dunlop Publisher Services Incorporated
Production Supervisor: Maureen P. Conway
Design and Illustration by Caliber Design Planning, Inc.

A special comprehensive Fingermath elementary school curriculum
is available to professionals. Kindly address inquiries to:

Fingermath International
School Program
McGraw-Hill Book Company
1221 Avenue of the Americas
New York, New York 10020

Library of Congress Cataloging in Publication Data

Lieberthal, Edwin M
    The complete book of Fingermath.

    1. Arithmetic—1961–   I. Title.   II. Title:
Fingermath.
QA115.L53   513   79-490

The Complete Book of Fingermath

ISBN 0-07-037681-6 (school)
ISBN 0-07-037680-8

Acknowledgments

"Everyone eats and drinks . . . few distinguish the flavors."

These simple words spoken by Confucius centuries ago still carry volumes of meaning. They can be applied to my own venture into Fingermath.

From the outset, it had been vividly clear that the nation suffered from a number of deficiencies, among them a widespread inadequacy in mathematics. Everyone knew it, lay persons and professional educators alike. Yet in spite of a glimmer of hope that this ancient remedy might help alleviate computational illiteracy, established habits and attitudes were to die hard.

There were a few, however, who early revealed themselves by openly withdrawing from the crowd and treading where others did not dare. Now, of course, the walls of resistance are crumbling, but at the outset these were the individualists who did, indeed, "distinguish the flavors" and in various ways lent methodical support to my madness.

Dr. Bruce Vogeli, Chairman of the Mathematics Department, Teachers College, Columbia University and Dr. Robert Postman of Mercy College, Dobbs Ferry, N.Y. By being learned devil's advocates, they helped me to distinguish passionate hopes from realities.

Dr. Leo Gafney, Senior Editor, McGraw-Hill Webster Division, helped me distill the essence of the method. Asked how Fingermath compared with the electronic calculator, Gafney observed, "It's the difference between playing piano and listening to a record of someone playing. One hears the music either way . . . but what a difference!"

Joseph Vellone, a true friend, gave never-ending support through many a critical day and helped me to maintain my equilibrium. (They have thrown away his mold.)

Arlene, Sandy, and Randi Wolf must be cited for always being there at the right time with compassion and understanding, above and beyond the call.

My daughters, Joy and Aimee, opened a new universe for my wife and me when they arrived from Korea two years ago. It was these little packages of sunshine and love that made us aware of the then dormant dynamics of finger mathematics. "Out of the mouth of babes...."

Bernadette, my perpetual bride, is—when all is said and done—the only wonder of my world. So together is this woman that, when the pieces of our life periodically are fragmented, her cohesiveness cements them together again. Loving yet unyieldingly critical, her ever-present scrutiny and obsession with detail made this work become whole.

And the tens of thousands who wrote letters of encouragement and inspiration all contributed to the Fingermath phenomenon.

How to use this book

If you're a teacher

Fingermath provides a manual reinforcement of the traditional classroom curriculum in arithmetic. Experience in teaching any subject convinces most professionals that there is an important difference between learning a subject oneself and teaching it to others. Fingermath is a prime example.

Learning Fingermath techniques is similar to mastering a sport or a musical instrument. Regardless of one's age, profession, intellectual level, or ability, the learning sequence to be followed is the same for everyone. The most elementary aspects of the subject must be addressed and mastered before one moves to higher ground.

For this reason, the elementary material in this book is intended for adults and children, teachers and nonprofessionals. All must go through identical sequences and acquire the same skills if they wish to become proficient in the method.

On the other hand, teachers are likely to progress at a steadier rate because of their familiarity with various mathematical concepts (though manual facility will not necessarily benefit therefrom).

This book attempts to convey, in simple terms, the entire series of manipulations that must be learned in order to perform Fingermath well. With the explanations given, a parent should be able to guide a child through the whole sequence of finger activities. However, the task of doing so with one or two of one's own children, in the leisurely convenience of the home, is vastly less complex than that of the class-room teacher. Training a large group of children in Fingermath, relating this method to basal curriculum, working for limited periods each day, adapting to varying levels of ability and interest, all complicate the teacher's task. For this reason, a separate, comprehensive, carefully devised classroom curriculum has been designed for teachers.

But the prerequsite for Fingermath classroom activity is mastery by the teacher of the skills themselves. (One could hardly be an effective piano teacher without knowing and experiencing the skills of playing the instrument.) In that regard, this book thoroughly prepares teachers to handle the special Fingermath materials that have been carefully planned, skill by skill, for daily classroom activity.

If you're a parent

If you wish to teach Fingermath to a child, there are a few suggestions you should follow.

Learn the method yourself before trying to pass it along. Stay at least several pages ahead and discipline yourself to practice just as you will expect your child to do.

Go slow. Very slow. Because Fingermath is so simple, you'll be tempted to race through the book. But remember that it takes time for a child to feel confident and secure about each new manipulation. Moving ahead quickly may add new skills, but the earlier ones will suffer and can be forgotten or get rusty.

Work every day for a *short* time. It's far better to spend ten minutes a day on Fingermath than to set aside an hour each weekend. Daily practice is the key.

Review each day the skills already acquired. Don't let them be forgotten. Like practicing scales on a musical instrument, a daily review is the best warmup.

Follow the sequence outlined. Don't skip sections, because each one prepares the learner for the next. If your child is weak in a particular area of arithmetic, don't try to get into it without first covering *all* the information that comes before it.

Fingermath is not a replacement for classroom work in arithmetic in your child's school. It is designed to supplement that work and, among other advantages, remove the pressure that some children experience when they must memorize tables and other numerical facts.

Speed kills. Ability to perform calculations rapidly should be a side effect of much practice; it should not be a goal that a parent sets. Aim at accuracy and comprehension; speed, as you will see, comes naturally as a by-product.

Fingermath is not a spectator sport. Try to get more than one child involved so that the gamelike, competitive aspects can contribute to performance.

It's unnecessary for a child to write in this book. When you reach the sections with practice pages, give your child a separate sheet of paper to enter numbers. It's a little less convenient, but it will keep this book looking clean and fresh.

If you're a student

You're old enough and bright enough to use this book on your own, without adult supervision. You'll find very little to confuse you, because Fingermath is a simple method. However, there may be stumbling blocks if you don't take sufficient time to learn each skill thoroughly.

Learning Fingermath is similar to learning a sport or a musical instrument. It takes plenty of practice.

You'll be tempted to skip over some portions because you'll be eager to go to more favorite subjects. Don't do it. Every skill prepares you for those that follow, and the entire sequence has been very carefully devised.

Don't allow your pre-Fingermath knowledge of arithmetic to interfere. Go through even the most elementary calculations (like $1 + 1$) as though you had never seen them before.

Now, take a glance at the above instructions to parents because they make other suggestions that you will find helpful. Good luck!

If you're the teacher or parent of a Learning Disabled or Handicapped Person

Experience thus far has indicated that the Fingermath method has no age barriers, no language barriers and no intellectual barriers. Because of the many sensory involvements with Fingermath, many students with even severe handicaps (blind, deaf, motor) are performing well. The motivational factor appears to be at such a high level that involvement becomes very active, thus producing notable results.

The procedures for teaching the Learning Disabled would not vary significantly from what has been outlined for so-called "normal" individuals, except to stress the obvious need to proceed even more slowly. Be certain to recognize the mastery of every skill before moving on to the next. In addition, be especially careful to review established skills frequently.

Additional lessons specifically related to teaching Fingermath to the blind, the deaf and the multihandicapped are being researched and developed for use by professionals and parents.

Foreword

by Dr. William E. Lamon, Director, Psychological Research Laboratory for Mathematics Learning, University of Oregon

The educational community is aware more than ever of its obligation to prepare *all* children to function in the competitive technological society they will inherit. Yet more and more children emerge from our elementary schools as arithmetical cripples, handicapped by their inability to utilize the four basic arithmetical functions. They cannot add, subtract, multiply, or divide.

Still, success in learning arithmetic does not elude all children, given a host of teaching aids and an educational setting conducive to pleasurable and meaningful learning experiences. The difference between the achievers and the nonachievers seems to be related to a difficulty in sensing the abstract significance of operations performed with teaching aids.

In mathematics learning, sensory input is imperative. For the arithmetically deficient child, tactile manipulation is crucial. Then what better way of learning than with the use of one's own hands? The pressing and releasing of the fingers, as in Fingermath, seems to facilitate a smooth transition between sensory input and its mental interpretation.

A year of involvement in the education of both normal and obviously handicapped children has demonstrated to me that exposure to Fingermath techniques is helpful as well as enjoyable for the children. Those of us who teach elementary school do believe that those children who are successful in their learning of mathematics, develop a pervasive feeling of well-being that is an accumulation of many pleasurable associations with mathematics, and that such satisfaction usually determines an appetite for further mathematical experiences. It is not certain whether the benefits gained are carried over into higher mathematical education, when reliance on the fingers is no longer necessary. But at least the advantage is clear in the elementary grades in cases where traditional learning aids have been ineffective. As a supplement to current pedagogical techniques, Fingermath is highly recommendable.

Contents

Introduction

Life is full of surprises. No sooner do you get used to the way society expects you to act, setting out ground rules for what is and what is not acceptable behavior, and no sooner do you pass these hard-won virtues on to your loved ones, than something comes along to turn your whole universe upside down.

Yet, if one can become used to the thought that the only constant thing is change, such rude awakenings can be a joy. With that paradox in mind, I welcome you to the world of Fingermath. I can assure you that, before you're a third of the way through, you'll have devised several answers to the question that little children everywhere are asked: "Why do you count on your fingers?"

Did I say little children? I mustn't exclude most 17- or 27-year-olds. After my TV appearances on the "Today" show and shortly thereafter with Johnny Carson on his "Tonight" program, I held my breath, expecting viewers to write scorching letters about my claim that fingers can outperform the calculator in both accuracy and speed.

Earlier, in discussions with friends about the idea of offering finger mathematics to the public as a dynamic method of helping children and adults with arithmetic, I invariably was given the fisheye and an incredulous "Are you serious?" No wonder, then, that in the initial deluge of mail (nearly 300,000 letters) that arrived over the next few months, my wife and I were mystified and delighted to find not a single comment that was negative.

The word that did come through, loud and clear, was "Help!" Boys and girls, as well as adults in every profession and occupation, clamored for information about the method. Universities and colleges; private, public, and parochial schools; prisons and prisoners; institutions for the blind, the deaf, the multihandicapped, the learning disabled; parents of whiz kids and fizzled-out kids; impresarios and promoters; newspaper, magazine, radio, and TV journalists; personalities and publishers; doctors, lawyers, dentists, architects, realtors, chefs, teachers, surgeons, engineers, merchants, manufacturers, bank clerks and bank presidents; congressmen, judges, pilots; the bed-ridden, the hospitalized, and the aged. They all wrote.

Urgent as an SOS from the *Titanic* came the plea for help from every corner of the United States and later from many foreign countries.

The need was evident, yet we could not help believing that, though the TV pyrotechnics had captured everyone's attention for a moment, the age-old hang-ups would prevail.

Wouldn't most parents keep on admonishing their children not to count on their fingers?

Why is that? Do you, deep down or perhaps very openly, feel that way? Does it embarrass you to see someone play the piano, paint a portrait, type a letter, build a house, do needlework? Or press the buttons of an electronic calculator? If there is something primitive about using one's fingers, why is it acceptable in manual activities such as these? Can you imagine going through a day in your life without using your hands?

Why is it, then, that these same miraculous human instruments suddenly become objects of scorn and are hidden under the table when used for work with numbers?

Will you believe me when I tell you that I have never heard a single intelligent answer to that question?

There is none!

Your hands are a great natural resource and you should be proud to use them in any creative way.

You already use your fingers to work a calculator, don't you? So there's no reason to be embarrassed to use these same fingers in place of a calculator. By depending on your own natural assets, you will discover a new avenue to self-esteem and accomplishment.

Can you or your child do basic examples like these, accurately and with enjoyment?

86,592 43,784 96,858 + 70,623 49,459	864 X39	6,000,400 − 839,257	14) 29,684	23 + Y = 47

Would you like to know how to use the four basic operations of arithmetic to perform very special and sophisticated computations, like these?

9 □ 6 8 □ 2 4 5 3 7 □ 4 + 8 5 9 □ 2 9,0 5 2	72 ÷ 8 X 6 − 12 + 54 = □	− 2,965 8,314 − 1,782 5,897

My 8-year-old does all of these . . . and she's not unusual. She uses her fingers. Hundreds of other children are solving the same kinds of problems.

So can you and yours.

Fingermath is not a panacea. I cannot promise or even suggest that it will cure the computational illiteracy in the world. What it can do—and is doing—is give children a sense of security with numbers very early in their lives. Instead of growing up with the anxieties that many adults feel with math, they are experiencing the excitement and fulfillment of arithmetic, the enrichment that comes with success.

Look at your hands.

The fingers are connected to the wristbone, but they're also connected to the headbone! And through the pages of this book, that path from mind to fingers is going to open up a new world of numbers for you. Whether your own arithmetic is rusty or you're simply concerned about your child, if you have two hands available you're on your way.

You will not win the Nobel Prize in mathematics even if you become the best Fingermath manipulator in the world. (The fact of the matter is that no one will ever win that prize; there isn't one awarded.)

However, at least now there is something besides death and taxes you can really count on. Your fingers.

Preparing to Calculate

The only instruments required to present your new Fingermath ballet of numbers are your two hands.

If you spend only a few minutes to discover the numerical values represented by each finger, you will realize that what you have here is essentially a decimal system, pure and simple. I say pure because of the consistent way Fingermath works: once you learn the structure, it never changes. You can depend on it. It is simple, because there are no complicated mathematical formulas or concepts involved.

You'll start out with your RIGHT HAND, because UNITS (ones), always appear on the right. No matter when or where you learned arithmetic, you soon discovered that units belonged in the right column. Now all of these single-digit numbers, from 1 to 9, will always be represented by the fingers on your RIGHT HAND.

How can you represent nine numbers on only five fingers? Simple. Take a look.

99 Fingers

THE RIGHT HAND: UNITS

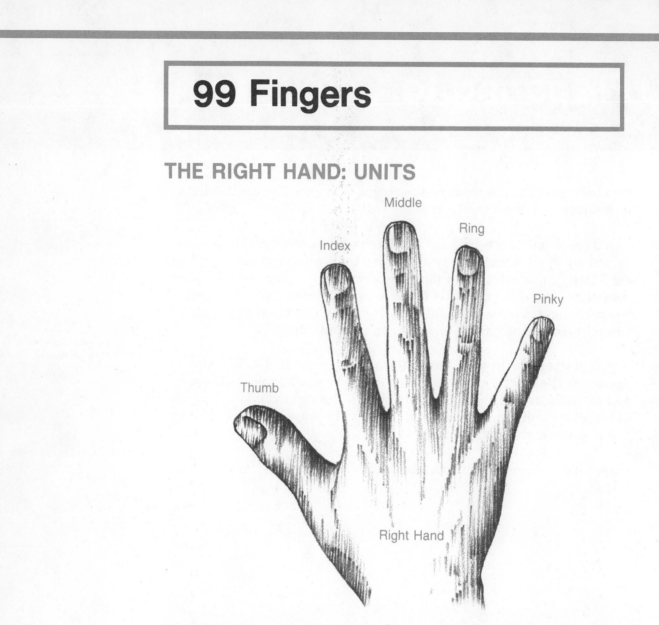

Note that every one of the fingers, except the thumb, has a value of 1. This means that whenever you use any of these fingers, it will always represent 1. Only 1. Always 1.

The RIGHT THUMB indicates 5, or five units. It always represents 5.

Therefore, it's easy to see that your right hand has a total value of 9. You might say you have the equivalent of nine fingers (five of them contained in your thumb and one each in the other four fingers).

As you'll see shortly, your right hand will always have the task of using its fingers to count, store, and calculate the numbers from 1 to 9.

THE LEFT HAND: TENS

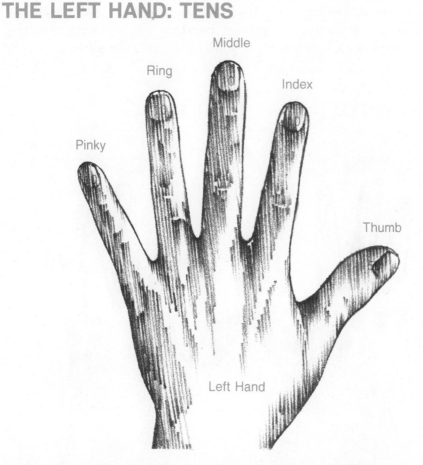

Middle

Ring

Index

Pinky

Thumb

Left Hand

In the same manner, every one of your LEFT HAND fingers, except the thumb, has a value of 10. This means that whenever you use any of these fingers, it will *always* represent 10. Only 10. Always 10.

The LEFT THUMB indicates 50, or five 10s. It *always* represents 50.

So, your left hand has a total value of 90. Once again, you have the equivalent of ninety fingers (fifty are in your thumb and ten are in each of your other four fingers).

Later on, you'll discover that your left hand will always be responsible for using its ninety fingers to count, store, and calculate all of the 10s from 10 to 90 (10, 20, 30, 40, 50, 60, 70, 80, 90).

And, just as you were always taught, the 10s are to the left of the 1s; your hands will see to it that they always stay there. It's something you'll never have to think about. If you're a southpaw, ignore it, because left-handed people do Fingermath in the same way as those who are right-handed. Like a typewriter, it works the same for all of us.

PUT BOTH HANDS TOGETHER

Fingermath gives you 99 fingers that will accumulate, store and calculate numbers

Now take a look at both hands side by side—just as you wear them. It's clear that each finger on the left hand has a value that is ten times greater than the corresponding finger on the right hand.

You no longer have ten fingers. You have ninety-nine. And you're going to make every finger count!

THE CORRECT FINGERMATH POSITION

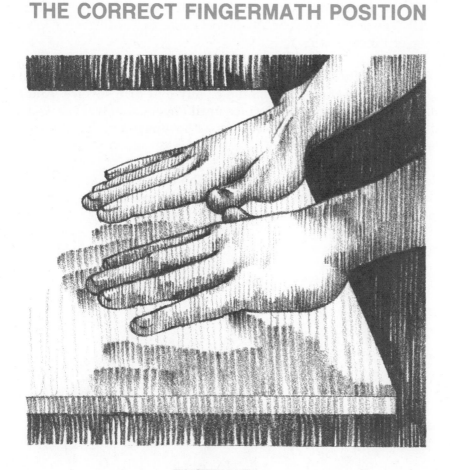

FINGERMATH

SEE IT.   FEEL IT.   SENSE IT.

Aside from its superb simplicity, one of the best features of Fingermath is that it makes use of your own natural human equipment. Instead of relying on some mechanical or electronic device, Fingermath uses only your mind and your body, one to learn and the other to perform the method. You will see, feel, say, and hear every Fingermath activity.

Your faculties will be so fully occupied that your attention will not be easily distracted. Children are captivated from the moment they begin, and adults—often skeptical at first about the potential of finger computation—soon become engrossed in the variety of sophisticated yet simple skills.

To make the most of this total involvement, your tools have to be used correctly. The position of your hands will make all the difference in your mastery of Fingermath.

I always think of the piano keyboard when I do Fingermath. Instead of keys, though, I need only a tabletop or any flat, firm surface. And my hands keep quite clear of the surface until I begin to play. I suspend my hands above the table, ready to press the imaginary keys with my fingers.

While there's no rigid rule for position, the drawing (page 9) shows the way to keep your arms over the surface. Relax your fingers in an open, spread fashion. Don't ever turn any of your fingers under like this:

NEVER CURL FINGERS UNDER LIKE THIS

PREPARING TO CALCULATE

Aside from the fact that it's tiring this way, your fingers won't be ready to act when they're needed. Think again about the keyboard of a piano or a typewriter. You'd never curl the idle fingers under while the others were in use.

Once your hands are suspended correctly, you're ready to strike any note simply by pressing down a particular finger or combination of fingers. That is, you will be pressing a number value by striking the surface.

A movement like this is called a finger manipulation. As you PRESS various numbers, you'll be able to feel the value as well as see it. In fact, as you progress through Fingermath, doing manipulations with your eyes closed will be an easy and frequent exercise. (This simply acquired habit of "seeing" with your fingers is one of the reasons why Fingermath can be learned so easily and used so effectively by the visually handicapped.)

HOW TO READ FINGERMATH DIAGRAMS AND INSTRUCTIONS

It's easy to follow the manipulations shown throughout this book if you understand the use of the special diagrams.

To give you the clearest direction for each manipulation, lifelike hand illustrations are used throughout the book in this way:

a) Since every calculation always begins with one or both hands in a suspended ready position, like this

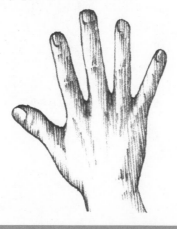

it is unnecessary to show this each time. Instead, hands will appear in the position of the first Press. For example, if instructions begin with the command "Press 1", the first illustration will appear like this. As an added visual aid, an orange dot appears above every finger that has been pressed.

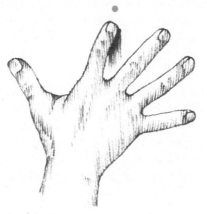

One finger is already established.

b) Every manipulation that follows the initial Press will be represented by an illustration in which the fingers are shown in the position just prior to the action. Directional arrows will indicate where each finger is to go during the manipulation . . . up or down. Therefore, you can study an illustration, place your fingers in the identical positions and then, to execute the action being discussed, follow the arrows. For example, if instructions begin with the command "Press 1"

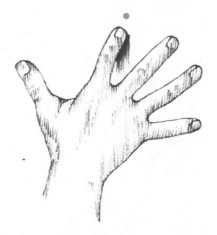

and the next command is
"Press One More",
the next illustration will appear like this . . .

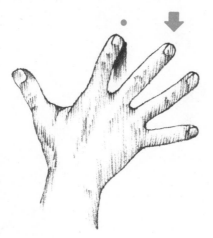

It shows that one finger is already pressed and one more should be pressed on that command.

c) The same method is used for Clearing. A command to Clear fingers will be accompanied by an illustration that shows where your fingers are before the manipulation and directional arrows indicating where each finger is to go during the manipulation. For example, if instructions begin with the command "Press 1"

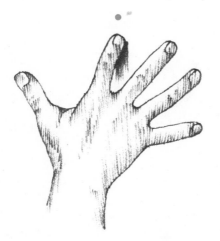

and the next command is "Clear 1", the next illustration will appear like this. It shows that one finger is already pressed and that this finger should be cleared on that command.

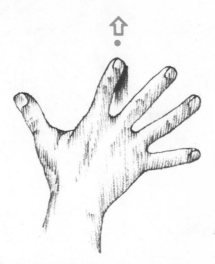

Therefore, every illustration, except the first, shows you the position of your fingers just before they act. Arrows, down or up, show you the direction in which any finger is to move to properly execute the command. Here's a final example of three commands in a typical series of manipulations.

A) **PRESS 1**

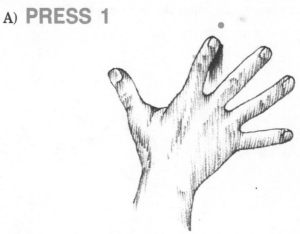

B) **PRESS 1 MORE**

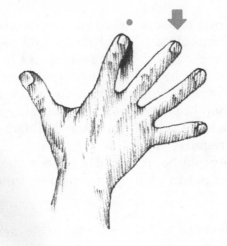

C) **CLEAR 1**

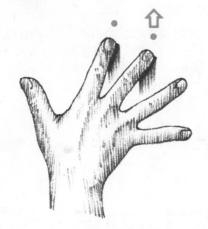

FINGER-PRESSING SEQUENCE

There is one final rule to learn before you get into action with Finger-math. And this is one place where this method parts company with piano-playing and typewriting. The new rule has to do with the order in which the fingers must be pressed. Playing the piano, you might press your thumb, your middle finger, and your pinky simultaneously to sound a chord. That just can't happen with Fingermath. The basic principle, which makes it possible for you to build up numbers correctly and scientifically, requires you to use your fingers in a carefully developed sequence. They may be combined only according to that sequence. In other words, you mustn't make up your own tunes or improvise in any way.

A similar rigid sequence is followed with other devices, such as the abacus, from which Fingermath was developed—or perhaps it was the other way around.

I think it's reasonable to assume that the first tools ever used to calculate numbers were fingers, and that the abacus was modeled on the hand, which, under the guidance of a wondrous brain, may be credited with the earliest developments of computation.

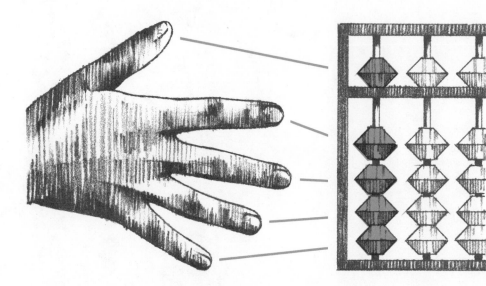

The ancient abacus probably was modeled after the hand. Note the similarity between the numerical values represented by each.

At any rate, there are no exceptions to the following Fingermath rule. It applies equally to both hands, even though they often function independently.

You must always press fingers in the order shown (A–B–C–D). You begin with the index finger, then the middle finger, next the ring finger, and last the pinky. (Forget about the thumb for the moment.)

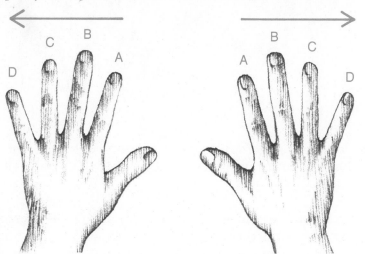

You may never skip a finger. In other words, you can never begin a manipulation by pressing your middle finger (B). Your index finger (A) must always be pressed first. It must remain pressed when you go on to press the middle finger. You will be accumulating pressed fingers, not replacing one with the next.

For example, you may have fingers pressed this way:

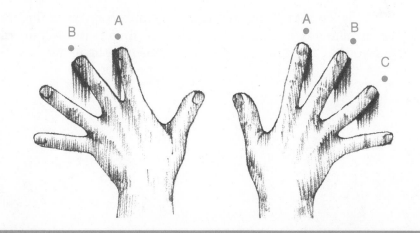

but you may never have them pressed like this:

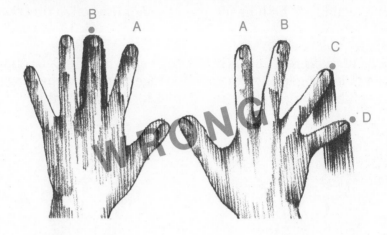

In order for these fingers to be pressed, you would first had to have pressed the left index finger (A) and the right index and right middle fingers (A and B).

In sum, when you count in sequence the only acceptable pressing order is INDEX FINGER–MIDDLE FINGER–RING FINGER–PINKY. You must never have the pinky pressed without the other three fingers. You must never have the ring finger pressed without the middle and index fingers. You must never have the middle finger pressed without the index finger. The index finger is the *only* one of the four fingers that may be pressed alone.

The thumb—I'll give special attention to it shortly—also may travel alone, as you shall see.

You will also learn later that, just as you establish or press fingers in fixed order from index to pinky, so do you clear fingers in reverse order. But let's save that for now and get to the business at hand.

A WARNING!

By this time, you're probably thinking you've had enough about sequences and hand positions and rules. The name of the book is Fingermath, so you have a right to expect some math. It's coming, I promise you.

But I have to caution you about jumping the gun. Even though the method is really simple, there's more to it than pressing the buttons on a calculator (although Fingermath can be more rewarding and much more fun). You just can't learn it right and do it well in one session—or in ten! The first thing you've got to believe is that the rules of the game are really worth your while to learn. So move slowly.

One other thing. Don't skip any section of this book. If your weak spot in arithmetic is subtraction, don't skip the sections that come before subtraction. Fingermath has to be learned in a carefully structured sequence. If you wanted to learn to dive and decided to skip "How to Swim," you'd really be in deep water. The same fate awaits you in Fingermath if you skip ahead.

ESTABLISHING NUMBERS: Right Hand

"When can a child begin to learn Fingermath?"

There is no specific answer to that question except to say that the only thing needed to start with Fingermath is the ability to read, write and say the numbers from 1 to 9. That basic knowledge and skill should be established. Age does not appear to be of any consequence; dramatic success has been noted with 5-year-olds. Neither does the ability to learn Fingermath depend on exceptional mental facility. There is some encouraging evidence that children previously thought to be learning disabled can perform well with this method.[*]

It's a great help if another member of the family or a friend can work

---

[*] In addition to his studies of the learning disabled, Dr. William E. Lamon, of the University of Oregon, is conducting a major research project to determine the best applications of finger mathematics for the handicapped. Early results encourage those involved to believe that the method offers a productive new avenue of learning for the blind, the deaf, and the multihandicapped.

with the learner, because many of the exercises in Fingermath require someone to call out instructions in order to provide practice in finger manipulation.

Facility in recognizing finger values without hesitation is among the most essential skills in Fingermath. It is an absolute must for Fingermath computations. All of the following section (through page 61) is aimed at helping you to recognize finger values. This is not the same thing as addition. It is, rather, a technique that enables you to do two things:

> First, to see or hear any number from 1 to 99 and respond by instantly pressing the finger or combination of fingers that represents that number.
> Second, to accumulate finger values almost automatically in later manipulations and to recognize instantly any desired predetermined value.

You'll hear the instruction "Clear your fingers" from time to time. This means you are to cancel out all previous manipulations and reset your hands in their READY position, suspended above the surface and prepared for fresh action, like this:

Clear Your Fingers. Then They're Ready To Begin A New Manipulation.

Now keep the book open to your left and work with your RIGHT HAND only.

FIRST, CLEAR YOUR FINGERS
(right hand READY)

(1) **PRESS 1**

(Say "One" as you press 1. Press
the index finger only; the other
fingers remain up.)

CLEAR YOUR FINGERS

(index finger up again in ready position)

Once again:
PRESS 1

(say "One")

CLEAR YOUR FINGERS

(all fingers up in ready position)

Now a step farther:

PRESS 1

(As you press, say "One")

PRESS 1 MORE

(say "Two" and be sure to keep the index finger pressed)

CLEAR YOUR FINGERS

(all fingers up)

Without fanfare, you've already gone through the following manipulations to arrive at a value of 2.

(2) **First, PRESS 1**

(say "One")

Then,

PRESS 1 MORE

(say "Two" and keep your index finger down)

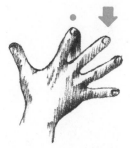

Perhaps you're thinking, "I'm way ahead of him. If he wants me to press 2, I'll just skip the finger-by-finger press and go direct to 2." But don't give in to the temptation. It's true that you already know how to press 2. But sticking with this basic-level counting will set you up for things you haven't yet anticipated. Take the time to crawl through these elementary exercises and you'll be zipping along later on. Remember the diver who couldn't swim!

Now CLEAR your fingers, ready for the next step. (Your fingers do the walking in this book too.)

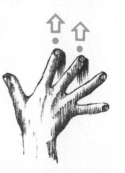

(3) **First, PRESS 1**

(say "One")

PRESS 1 MORE

(say "Two")

PRESS 1 MORE

(say "Three")

(Feel 3. Press hard.)

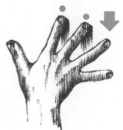

CLEAR YOUR FINGERS

(4) **First, PRESS 1**

(say "One")

PRESS 1 MORE

(say "Two")

PRESS 1 MORE

(say "Three")

PRESS 1 MORE

(say "Four")

CLEAR YOUR FINGERS

Once more:

PRESS 1

(say "One")

PRESS 1 MORE

(say "Two" and keep index finger down)

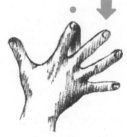

PRESS 1 MORE

(say "Three" and keep index and middle fingers down)

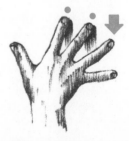

26

PRESS 1 MORE

(say "Four" and keep index, middle, and
ring fingers down)

Feel your fingers pressing. Sense what 4 feels like. Try to push your
fingers through the surface. Close your eyes, press harder and say
"Four." Make sure your thumb is still clear.

CLEAR YOUR FINGERS

to the ready position, and then let
them stand at ease for a time.

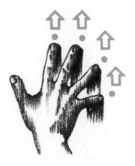

From this point on, I will be telling you to press not only 1, but at times
2, 3, or 4 or more. No matter what number I tell you to press, you will
reach that number by pressing one finger at a time, as you have been
doing. This is called unit-by-unit, or finger-by-finger, or one-by-one
progression.

You will go through several levels of technique in Fingermath. The
most basic is always restricted to this unit-by-unit count.

You will be using basic-level Fingermath for some time, so get used to
it and practice saying each number aloud with each press until it
becomes comfortable and automatic. Do not abandon this technique
until you are instructed to replace it with a more advanced technique.

THE RULE OF THUMB

You've already learned that in Fingermath the thumb has a value of 5 on the right hand and 50 on the left hand.

As you've just seen in pressing the numbers 1, 2, 3, and 4, you started counting when you pressed the right index finger (one); you continued with the middle finger (two); then the ring finger (three); and then the pinky on the count of four. Now, at the count of five, you will press your thumb. At the same instant, you will clear the first four fingers. This simultaneous manipulation, called an EXCHANGE, looks like this:

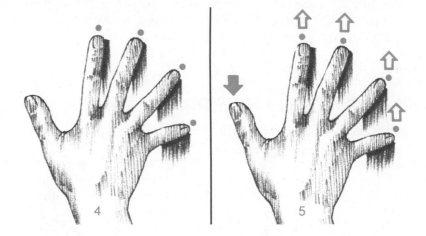

To count from four to five, simultaneously PRESS 5 (press the right thumb), say "five," and CLEAR all four right-hand fingers.

What you've done is EXCHANGE your first four fingers for your thumb. You've exchanged 4 for 5.

This thumb will work in exactly the same way on the left hand, when you go from a count of forty to fifty. But you're not ready to tackle the left hand yet, so when the time comes, I'll refresh you on this point.

Meanwhile, practice the right-hand EXCHANGE from 4 to 5, counting aloud. Start with a press of 1 and continue unit-by-unit to count up to five. Remember, whenever your first four fingers are pressed and you want to add 1 more, you *must* do this exchange. If you simply pressed

28

your thumb and did not clear the other four fingers, your entire hand would be pressed ... and that would give you 9.

You can see now that your thumb does two things:

first, it holds the position of the fifth unit, when you are counting by units;
and second, it represents the value of the first four units plus its own 1, for a total numerical value of 5.

NOW HERE IS THE FINGERMATH RULE OF THUMB:

To add ONE to a press of 4, you must simultaneously press the thumb and clear the established four fingers, releasing them to be used again as new units in combination with the thumb.

Because your thumb stands out naturally from the other four fingers on your hand, this 4 to 5 exchange is simplified. But you've got to practice it, starting with 1. Do it now, at least ten times, before you go on to the next manipulation.

On the instruction "Press 5":

⑤ First, PRESS 1

(say "One")

PRESS 1 MORE

(say "Two")

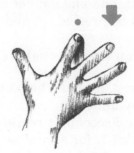

PRESS 1 MORE

(say "Three")

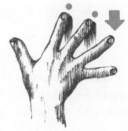

PRESS 1 MORE

(say "Four")

(say "Five" and simultaneously make the exchange)

When you execute this 4 to 5 exchange, do it sharply and cleanly. Don't allow yourself to crawl through it. This is not a two-step manipulation. You must not press your thumb and then a second later release your other fingers; these operations must occur at the same instant. As you say "five," simultaneously slap down your thumb and flip up your other fingers. For the time being, exaggerate the action. Overemphasize it. This 4 to 5 exchange will become one of your most frequent and most useful manipulations in the days ahead.

Be sure to practice pressing the numbers from 1 to 5. Do it at least ten times in a row, several times a day. Count aloud as you press each finger.

On the command "Press 5," you'll be tempted to go direct to your thumb for an instant press of the number 5. And you would not be wrong. Except that you would be avoiding one of the most necessary exchange exercises in the whole Fingermath method. At this stage, you must get to 5 by travelling the unit-by-unit, basic-level road. Very soon, you'll be introduced to the next level and you'll start pressing 5s direct. Don't cheat yourself. Count it out, one by one. Ten times!

> Take note that the exchange from 4 to 5 is regarded as one of the few real devils of Fingermath. If you don't believe me, just wait and see. Or practice it faithfully every day and beat the devil!

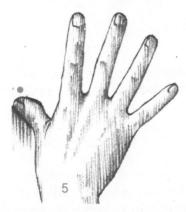

Now that you're on your way to clearing away the hazard of this 4 to 5 exchange, it must be obvious that you possess a mechanism like a calculator in your fingers. By way of this simple, instant transfer of values from the first four unit fingers to your thumb, you have set up the means for counting or accumulating still more value with the same fingers. Press your thumb (5) and observe your other, cleared fingers. They are free. Available. Ready to be used again. And they *will* be used again, in the same way you originally used them, as units. But this time you will combine them one by one with your thumb.

Before you do, though, this is a good place to tell you about the various levels that you'll be achieving in Fingermath. Although they all reach the same numerical conclusion, each level provides a more direct and rapid route than the one that preceded it. Thus each Fingermath level will introduce a major shortcut, a finger manipulation that bypasses some or all of the steps in the previous level.

You are ready now to get started with intermediate manipulations. If you're teaching Fingermath to a child, it's a good idea to continue at the basic level for a much longer time. This basic unit-by-unit counting can be used successfully at any time with all or any of the calculations, and you'll be surprised to find yourself going back to it now and then, whenever the more advanced manipulations give you trouble. Like an old shoe, it always feels comfortable and takes you where you want to go.

Like a pianist who takes time to practice the most basic, finger-by-finger exercises *every day,* you will develop dependable, automatic skills with Fingermath if you make it your business to practice your finger-by-finger, one-by-one manipulations regularly. The true virtuoso never gets too good to practice scales; neither should you.

INTERMEDIATE LEVEL

Just as basic-level Fingermath is defined as finger-by-finger, unit-by-unit counting, so manipulations at the next level can be described as follows:

> To press a value of 5 or more, you will first press the 5 (the right-hand thumb) directly. Then, if needed, you will add to that via the basic level—by units. So, strictly speaking, the intermediate level uses some basic techniques. You will find this true of all higher levels: they always continue to employ some of the techniques from levels you've already passed. This is the main reason you should never pass over any portions of this book. By skipping ahead, you would most certainly miss learning some very essential maneuvers.

Following are instructions for pressing the numbers 5, 6, 7, 8 and 9,

using your new intermediate technique. Go through each step, saying the numbers as indicated. Then go back and try each one using the unit-by-unit, basic method, beginning always with the count of one. This is not a step backward for you. It serves to reinforce your 4 to 5 exchange, for one thing, and it emphasizes the way your fingers work both independently and in concert with your thumb.

Finally, close your eyes and feel your way through the pressing of numbers 5 through 9, first using intermediate and then basic techniques.

Now we can continue our progress beyond 4.

(5) **PRESS 5** direct (say "Five")

(You have bypassed the basic-level count-ing of one–two–three–four and arrived at 5 with a single press.)

CLEAR YOUR FINGERS

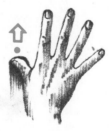

(6) First, **PRESS 5**

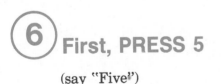

(say "Five")

PRESS 1 MORE

(say "Six")

CLEAR YOUR FINGERS

(7) **First, PRESS 5**

(say "Five")

PRESS 1 MORE

(say "Six" and keep your thumb pressed)

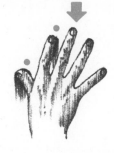

PRESS 1 MORE

(say "Seven")

CLEAR YOUR FINGERS

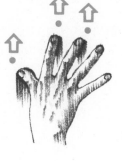

(8) First, PRESS 5

(say "Five")

PRESS 1 MORE

(say "Six")

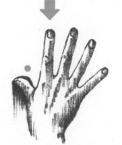

PRESS 1 MORE

(say "Seven")

PRESS 1 MORE

(say "Eight")

CLEAR YOUR FINGERS

(9) **First, PRESS 5**

(say "Five")

36

PRESS 1 MORE

(say "Six")

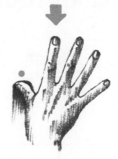

PRESS 1 MORE

(say "Seven")

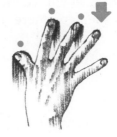

PRESS 1 MORE

(say "Eight")

PRESS 1 MORE

(say "Nine")

CLEAR YOUR FINGERS

STOP   Go back now and repeat each of these as I outlined earlier. Practice them unit-by-unit from 1, and then practice them again by the intermediate route. Repeat the same exercises with your eyes closed. Don't go ahead until you do.

Look again at the position of your right hand as you go through these maneuvers. Are your fingers spread comfortably? Do your unpressed fingers remain well clear of the surface, ready for action? Are you performing the 4 to 5 exchange sharply, with a slap of your thumb and a simultaneous flip up of your other fingers? Are you saying each number aloud? Doing this connects the physical experience of pressing to the expression of that value. You've got to feel in your fingers the difference between 3, for example, and 8. By talking yourself through each exercise, this feel will be firmly established. (If you're embarrassed to count aloud, then go where nobody will hear. Only get it done!)

PRESSING WHOLE NUMBERS

It's a simple step from where you've just been to the next major Finger-math skill of pressing a whole number. There's a difference, of course, between counting up to 7, say, and pressing 7 direct. Until now, whether you counted by units (basic) or began with 5 (intermediate), you have reached 7 in several steps. In order to reach 7 in a single action, you must be able to respond to the command "Press 7" or to the written number 7 by sensing in your fingers what that sound or symbol feels like when pressed. You've got to fix that feeling in your memory so clearly that when you want to establish a 7, your fingers will do it for you automatically.

Before very long, if you practice only a short while each day, you'll be establishing each of the numbers from 1 to 99 with an instant, single press of fingers.

What's more, you will become capable of pressing any assortment of whole numbers in any sequence, *accumulating* their values. In other words, you'll have reached the math in Fingermath: you will be doing addition. And you'll never lose your place because your fingers, like a calculator, will store subtotals for you. Most remarkable of all, you will achieve superb accuracy at speeds that outperform the calculator.

With these incentives in mind, let me get you back on the subject of pressing whole numbers. Logically, we have now reached the

UPPER LEVEL

Since you've only learned to finger-count up to 9, your first attempt at pressing whole numbers should be in the familiar range of 1 to 9, where you are already expert with both basic and intermediate techniques.

From time to time I have suggested that you work with someone else. Though arithmetic is an activity you'll actually do by yourself, this particular phase of Fingermath involves so much learning on the physical, manipulative level that it takes on features often associated with sports. It's certainly possible to practice a good tennis swing by hitting the ball against a wall and then continuing to return it, solo. But any tennis buff will be quick to point out the advantages of practicing with another player. To begin with, there's the psychological effect of having another person observe your performance. In addition, competition tends to bring out the best possible execution.

For a child, the parent's participation is especially rewarding. There is a sense of pride in being able to equal or beat the efforts of an adult.

I feel, therefore, that if there are one or two other Fingermath enthusiasts to practice with, no matter what their ages, if they can keep hitting the ball back to you at a varied pace and in an unplanned sequence, all of you will benefit greatly. Playing tennis, you can never anticipate what the next stroke by your opponent will be and your reflex response is speeded up. Playing alone, against a wall, you can control the return and become lazy.

Now if you have someone to work with you, ask your partner to call out the following instructions, at first spacing them about five seconds apart. Then repeat the list over and over, speeding up gradually until you can instantly press any number given, whether your eyes are open or closed. Don't allow yourself to press in steps. When you hear a number, make sure you think it through for a single, firm press of the whole number.

> PRESS 2.   CLEAR your fingers.
> PRESS 6.   CLEAR your fingers.
> PRESS 5.   CLEAR your fingers.
> PRESS 3.   CLEAR your fingers.
> PRESS 8.   CLEAR your fingers.
> PRESS 4.   CLEAR your fingers.
> PRESS 9.   CLEAR your fingers.
> PRESS 7.   CLEAR your fingers.
> PRESS 1.   CLEAR your fingers.

Repeat the entire list several times, changing the order repeatedly.

This instant recognition of finger values is the most important Fingermath skill you can master, particularly with the numbers 1 through 9. And here's why.

If variety is the spice of life, then consistency is the bedrock. And if there's one single aspect of Fingermath that makes it simple to comprehend, it is its consistency. Once you learn to do something one way, you will always do it the same way. With this method, there's no such rule

as *i* before *e* except after *c*. It's this fact that makes Fingermath mathematically sound. It is not a bag of tricks that work in some situations and not in others. The rules, combinations, and techniques it employs are used consistently whether you're doing addition, multiplication, subtraction, or division.

What has this got to do with pressing whole numbers? Everything. As soon as you master the nine single-digit numbers, pressing them instantly and understanding how they work, you'll be prepared to use these fundamental manipulations for every one of the four operations of arithmetic.

Recognizing Right Hand Finger Values

Instructions: First, press what you see on the drawing . . . then select the correct number.

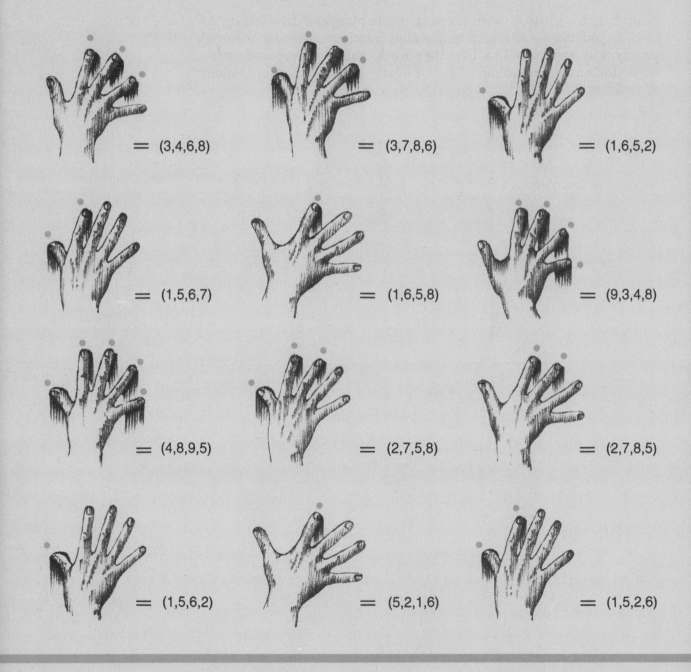

= (3,4,6,8)   = (3,7,8,6)   = (1,6,5,2)

= (1,5,6,7)   = (1,6,5,8)   = (9,3,4,8)

= (4,8,9,5)   = (2,7,5,8)   = (2,7,8,5)

= (1,5,6,2)   = (5,2,1,6)   = (1,5,2,6)

PREPARING TO CALCULATE

Child's Play

Press the value that appears in the circle. Then, with your left hand, point to the facts around it that correspond.

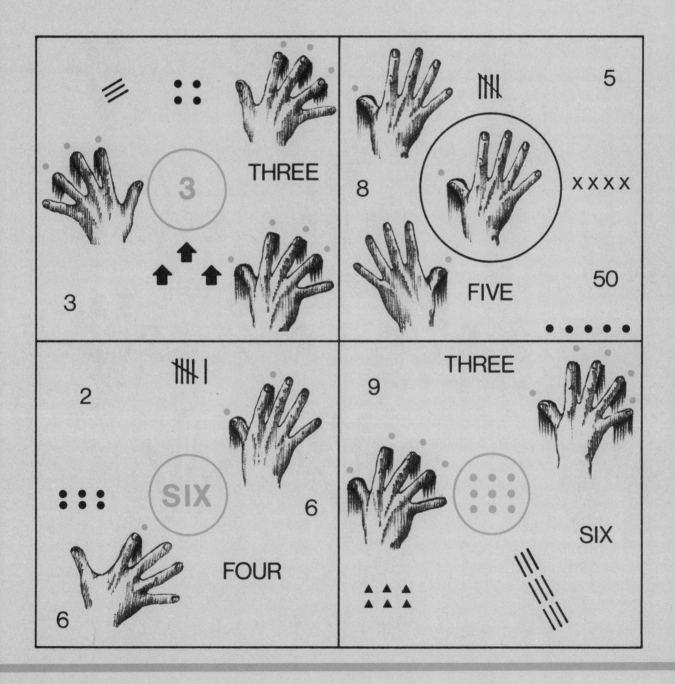

Press each number that appears in the left column. Then find that value on one of the diagrams below. Repeat several times. Then reverse the exercise. First press the hand value in each diagram and find the corresponding number:

8
4
5
2
7
0
3
9
1
6
5
8
7
2

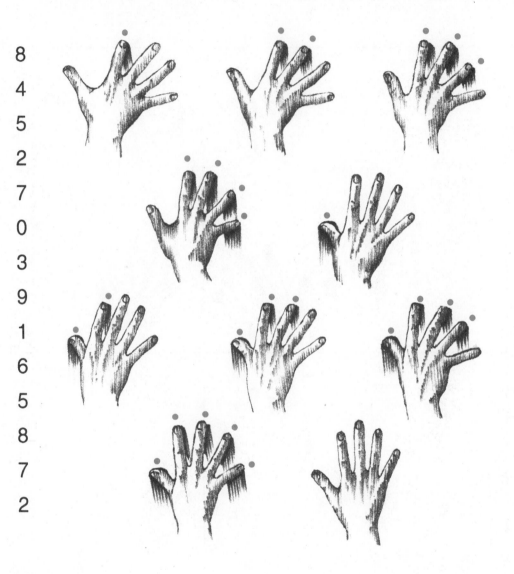

PREPARING TO CALCULATE

ESTABLISHING NUMBERS: LEFT HAND

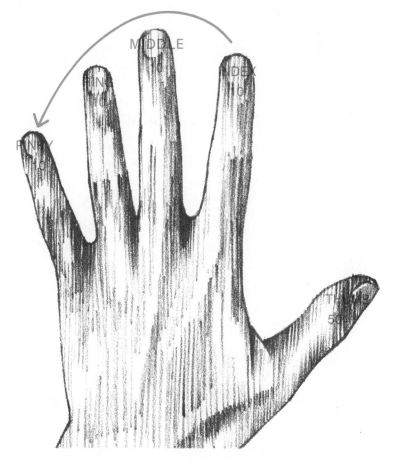

Now you've more than earned the privilege of getting your left hand into the action.

As essential as the right hand, your left will help you to finger your way through all kinds of sophisticated manipulations.

Because you already understand the right-hand structure and finger sequence so well, you'll spend on the left hand a fraction of the time you spent learning how to work with your right.

Each left finger has a numerical value ten times greater than its right counterpart. But its basic movements are identical. You still press your index finger first, then your middle, your ring, and finally your pinky. Instead of a 4 to 5 exchange, you'll be using the rule of thumb to do a 40 to 50 exchange. Then, again in the familiar right-hand pattern, you'll combine the newly cleared 10s fingers with the thumb to give you a total left-hand value of 90.

So get your left hand into a ready position and run through this finger-by-finger exercise. First, follow the diagrams and speak out as indicated. Then do it again with your eyes closed.

FIRST, CLEAR YOUR FINGERS

 **PRESS 10**

(Press your index finger only.)

(say "Ten" as you press 10)

CLEAR YOUR FINGERS

 **PRESS 10**

(say "Ten")

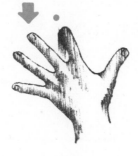

PRESS 10 MORE

(say "Twenty")

CLEAR YOUR FINGERS

(30) **PRESS 10**

(say "Ten")

PRESS 10 MORE

(say "Twenty")

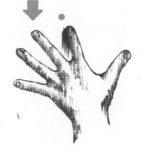

PRESS 10 MORE

(say "Thirty")

CLEAR YOUR FINGERS

(40) PRESS 10

(say "Ten")

PRESS 10 MORE

(say "Twenty")

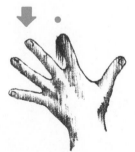

PRESS 10 MORE

(say "Thirty")

PRESS 10 MORE

(say "Forty")

CLEAR YOUR FINGERS

As you go through this exercise, you may be wondering which Finger-math level it belongs to. After all, you're pressing only one finger at a time (a basic technique), yet you're establishing a whole number, 10, each time you press (an upper-level technique). Well, take your choice of labels, because they don't matter. The only important consideration here is to observe the same rules of position and sequence as you did for your right hand.

At this point, you're ready to employ the rule of thumb again. The exchange from 40 to 50 has no surprises for *you*. You're still going to press your thumb as you simultaneously clear your four fingers. Try it out—by the numbers.

(50) PRESS10

(say "Ten")

PRESS 10 MORE

(say "Twenty")

PRESS 10 MORE

(say "Thirty")

PRESS 10 MORE

(say "Forty")

50

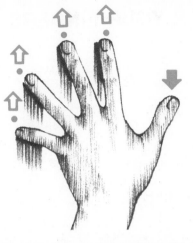

PRESS 10 MORE

To count by tens from 40 to 50, simultaneously PRESS 50 (press the left thumb), say "fifty," and CLEAR all four left-hand fingers.

YOU HAVE EXCHANGED 40 FOR 50

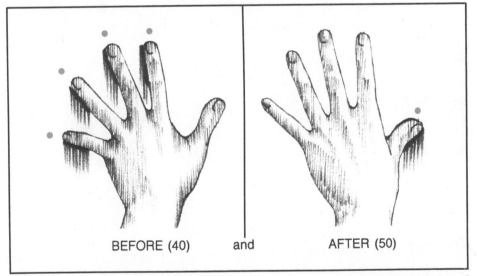

BEFORE (40)   and   AFTER (50)

I know you can easily find your way through the rest of the left-hand 10s, but I'd like you to have the whole sequence outlined for proper practice. Also, if you're teaching these techniques to a child, it's much easier with diagrams. Again, even for yourself, don't skip over what appears to be an obvious manipulation.

Continue, then, to press all the 10s by going the finger-by-finger route. This will take you through each preceding value several times.

CLEAR YOUR FINGERS

(60) **PRESS 10**

(say "Ten")

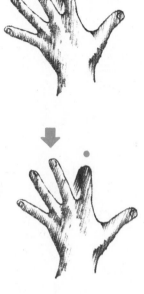

PRESS 10 MORE

(say "Twenty")

PRESS 10 MORE

(say "Thirty")

PRESS 10 MORE

(say "Forty")

PRESS 10 MORE

(say "Fifty")

PRESS 10 MORE

(say "Sixty")

CLEAR YOUR FINGERS

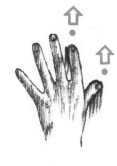

(70) **PRESS 10**

(say "Ten")

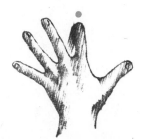

PRESS 10 MORE

(say "Twenty")

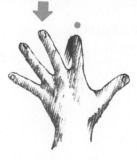

PRESS 10 MORE

(say "Thirty")

PRESS 10 MORE

(say "Forty")

PRESS 10 MORE

(say "Fifty")

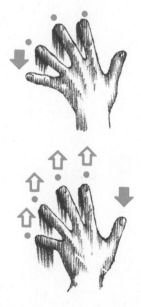

54

PRESS 10 MORE

(say "Sixty")

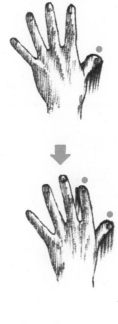

PRESS 10 MORE

(say "Seventy")

CLEAR YOUR FINGERS

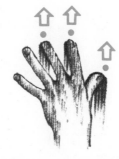

80 PRESS 10

(say "Ten")

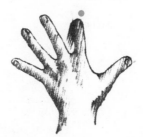

PRESS 10 MORE

(say "Twenty")

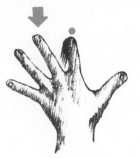

PRESS 10 MORE

(say "Thirty")

PRESS 10 MORE

(say "Forty")

PRESS 10 MORE

(say "Fifty")

PREPARING TO CALCULATE

PRESS 10 MORE

(say "Sixty")

PRESS 10 MORE

(say "Seventy")

PRESS 10 MORE

(say "Eighty")

CLEAR YOUR FINGERS

PRESS 10

(say "Ten")

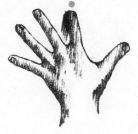

PRESS 10 MORE

(say "Twenty")

PRESS 10 MORE

(say "Thirty")

PRESS 10 MORE

(say "Forty")

PRESS 10 MORE

(say "Fifty")

PRESS 10 MORE

(say "Sixty")

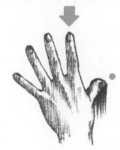

PRESS 10 MORE

(say "Seventy")

PRESS 10 MORE

(say "Eighty")

PRESS 10 MORE

(say "Ninety")

CLEAR YOUR FINGERS

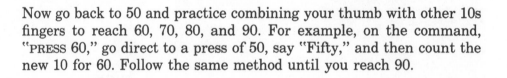

Now go back to 50 and practice combining your thumb with other 10s fingers to reach 60, 70, 80, and 90. For example, on the command, "PRESS 60," go direct to a press of 50, say "Fifty," and then count the new 10 for 60. Follow the same method until you reach 90.

Recognizing Left Hand Finger Values

Instructions: First, press what you see on the drawing . . . then select the correct number.

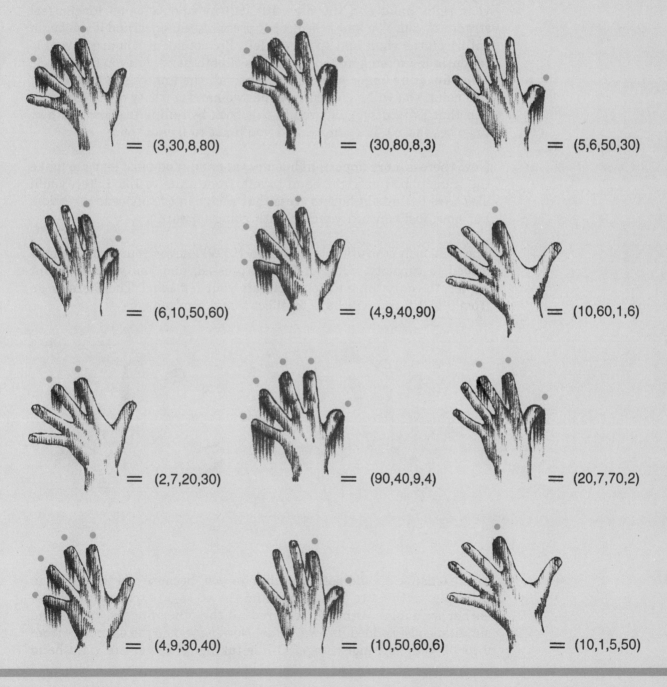

= (3,30,8,80)   = (30,80,8,3)   = (5,6,50,30)

= (6,10,50,60)   = (4,9,40,90)   = (10,60,1,6)

= (2,7,20,30)   = (90,40,9,4)   = (20,7,70,2)

= (4,9,30,40)   = (10,50,60,6)   = (10,1,5,50)

THE BRIDGE FROM UNITS TO TENS

Now you fully understand the numerical structure and the numerical capacity of both hands. So the tools of Fingermath—the nine units in your right hand and the nine 10s (90) in your left—no longer feel strange to you. You know the proper pressing sequence, and if you begin with 1 on the right and 10 on the left, I'm confident you are now able to continue pressing matching fingers of both hands (eleven at a time!), using finger-by-finger progression, until all the fingers are pressed. At this point, you will have reached the two-hand capacity of 99. (Have no fear that I will disappoint you in this book by failing to show you how to go beyond 99. It's simple, and you'll get to it later on.)

Now, there is more important business at hand. You must learn to make the transition from right hand to left, from units to 10s. Later, you'll also have to find out how to cross that bridge in the opposite direction. For now, just concern yourself with going forward.

The first such transition is from 9 to 10. Whenever your right hand is loaded to capacity, with all fingers pressed, and you want to count higher, the only thing to do is activate your left hand. Therefore, to go from 9 to 10, you will have to effect a major exchange:

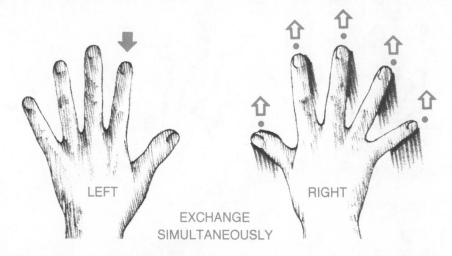

LEFT   RIGHT

EXCHANGE
SIMULTANEOUSLY

This exchange should seem familiar to you, because it's the same, in principle, as the 4 to 5 exchange and the 40 to 50 exchange. In those earlier exchanges a thumb acquired all the other fingers' values and simultaneously added its own value. Now, as you do the exchange from 9 to 10, your left index finger (10) is taking on the entire right-hand

PREPARING TO CALCULATE

value of 9 and simultaneously adding its own value. In one simple manipulation you exchange 9 for 10. Try it, following these instructions:

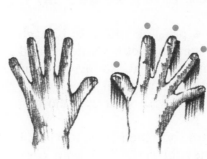

PRESS 9

(say "Nine")

PRESS 1 MORE

(say "Ten")

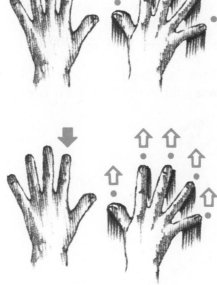

Be sure you execute a *simultaneous* exchange. Overdo it at first. Exaggerate the action by throwing your right hand up over your head. This overemphasis will ensure that you don't leave your right hand pressed. And there would be a big difference if you did, because your hands would look like this:

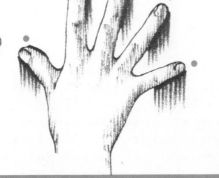

Instead of 10, you'd register 19. No good! Naturally, as you see, the proper exchange from 9 to 10 releases all the fingers of your right hand so that they can be used again. They must be clear, in a ready position, prepared to press again any value from 1 to 9.

Now practice your 9 to 10 exchange (it conceals another lurking devil) by starting your count at 1 and going forward at increasing speed each time you try it. Go through this exercise by units until you manipulate, without a moment's hesitation, the numbers from 1 to 10. Again, learn to do it with your eyes closed and always say the numbers aloud.

Once you reach 10, your right-hand fingers have the All Clear. They're ready for action and your count can continue.

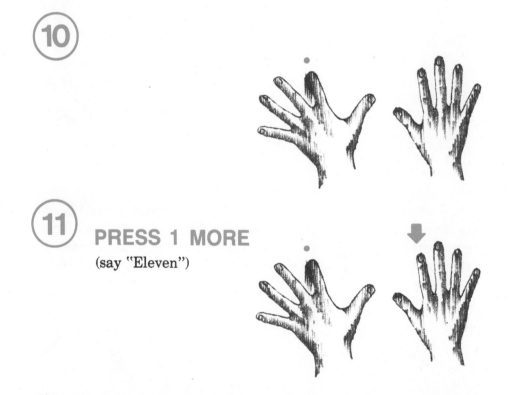

(10)

(11) **PRESS 1 MORE**
(say "Eleven")

This press even *looks* like 11. It's the first time you have had fingers of both hands pressed at once. Just a word here if you're teaching a child. Up to this point, the only thing needed for learning Fingermath has been an ability to read, write, and say the numbers from 1 to 9 in sequence (and perhaps count by 10s to 90). Now a new plateau is reached, running from 11 through 20. Be certain that these intermediate numbers can be read, written, and spoken in sequence before the child is expected to proceed with finger recognition of these numbers.

If your child is very young and numbers beyond 10 are too great a challenge, don't fret. Skip this section, wherein numbers are established to 99, and proceed directly to simple addition (page 89), and stay within the range of 1 to 10. Be certain first, however, that any whole numbers can instantly be pressed on commands given at random.

But don't skip anything yourself! Continue now with the numbers 12 to 19 by simply keeping your left index finger pressed (10), and pressing one unit at a time in proper sequence with your right hand. Count aloud as you take on each new unit.

I want to emphasize that you are not yet performing the operation of addition, even though, as an adult, you may perceive some resemblance to it in this accumulation of units. A child does not have this advantage of a head start in math (it's really a disadvantage, as you shall see), and goes through this procedure strictly to become intimate with finger values. The youngsters (and you) must come to recognize instantly whatever is pressed. Therefore, don't yet look at left-hand 10 and right-hand 1 as addition with a sum of 11. Rather, try to see and feel that combination of pressed fingers only as the whole number 11. There is a fine distinction here, but it's important to recognize it. Though a child may "read" a press of 16, for example, that ability does not include a knowledge of how many different combinations of numbers can be added to total 16. And there's the difference between number recognition and addition.

Having pressed 11, proceed in the familiar way with the next sequence of numbers.

(12) **PRESS 1 MORE**
(say "Twelve")

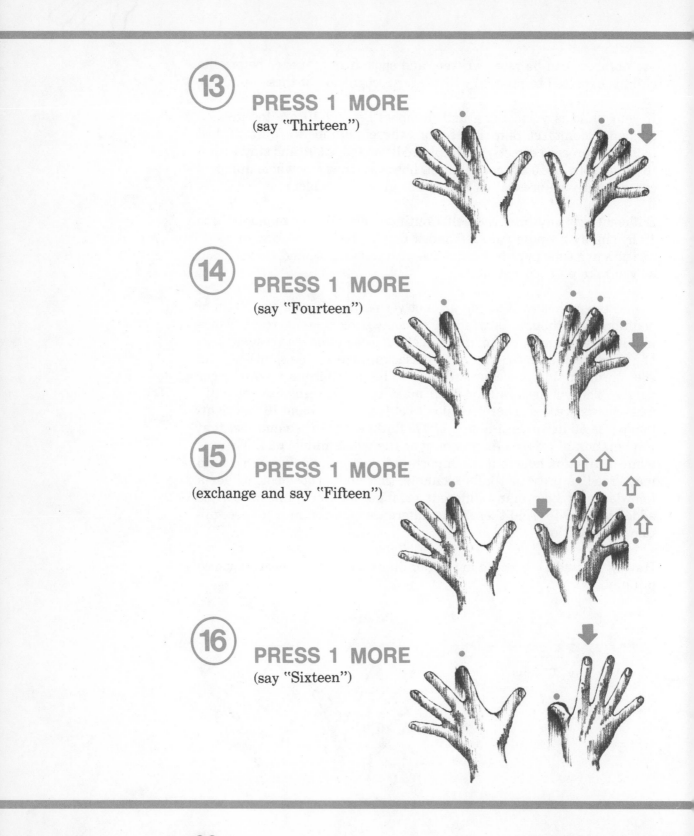

(13) PRESS 1 MORE
(say "Thirteen")

(14) PRESS 1 MORE
(say "Fourteen")

(15) PRESS 1 MORE
(exchange and say "Fifteen")

(16) PRESS 1 MORE
(say "Sixteen")

PREPARING TO CALCULATE

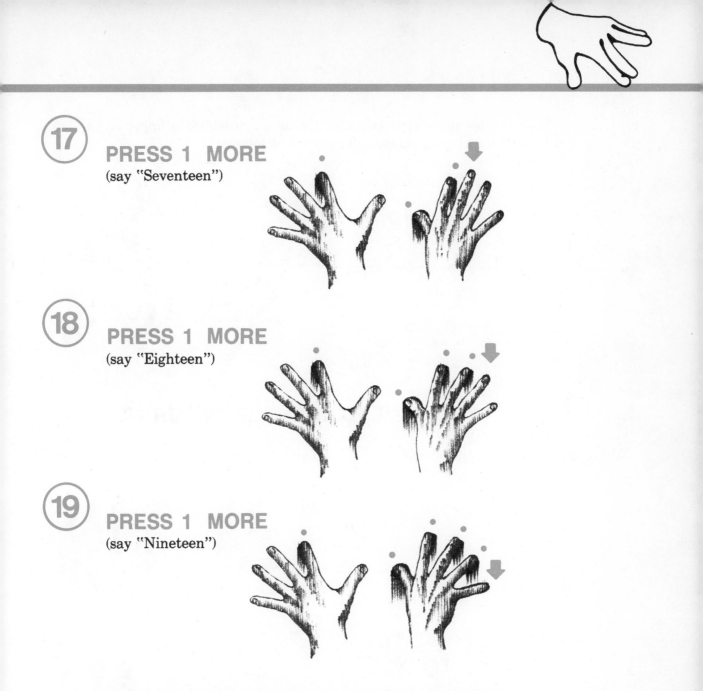

17 **PRESS 1 MORE**
(say "Seventeen")

18 **PRESS 1 MORE**
(say "Eighteen")

19 **PRESS 1 MORE**
(say "Nineteen")

Now keep 19 pressed. Look at it and ask yourself: "Where do I go from here to add the next unit?" Go ahead and try it.

That's good! Read on . . .

You can see that all transitions from units to 10s employ the same manipulation that you learned for going from 9 to 10. As soon as your right hand has no more fingers available, you will simultaneously exchange that entire hand for the next available 10s finger on the left hand. Having pressed 19, proceed as follows:

to

PRESS 1 MORE

"Twenty

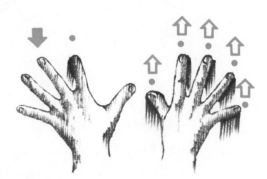

YOU HAVE EXCHANGED 19 FOR 20

Now count to 29.

to

PRESS 1 MORE

(say "Thirty")

YOU HAVE EXCHANGED 29 FOR 30

68

Now count to 39.

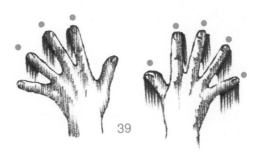

39

PRESS 1 MORE

(say "Forty")

YOU HAVE EXCHANGED 39 FOR 40

Do not proceed. Instead, take the time to press all the numbers from 1 through 40, in sequence. Continue this essential exercise until you can do it at a smooth, even pace, never hesitating for an exchange. This is an excellent group activity, with everybody calling out the numbers in unison. Go slow at first, then pick up more speed for each complete run-through. When you are quite sure of your own (or your child's) command of this sequence, you are ready to move forward.

WAIT

Now count to 49.

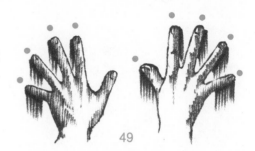

49

For this progression, you're faced with a somewhat more intricate exchange. The principle behind it is the same as with the others you have just performed. You have to exchange 9 for 10 by pressing your next available 10s finger (your thumb, correct?) as you simultaneously clear your entire right hand. Here it is:

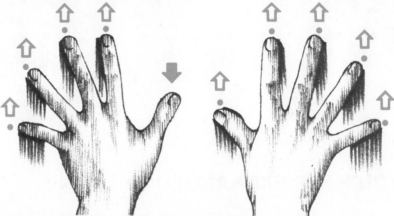

YOU HAVE EXCHANGED 49 FOR 50

Start again from 1 and practice counting to 50 without hesitation. Do it with your eyes closed, always counting aloud. When you feel that your fingers know where to go in this series, move on:

Count to 59

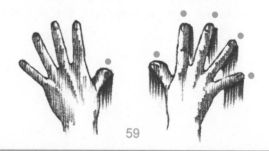

59

 to

PRESS 1 MORE

(say "Sixty")

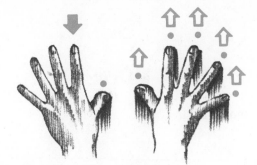

YOU HAVE EXCHANGED 59 FOR 60

COUNT to 69

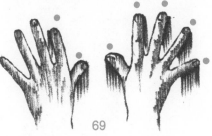

69

 to

PRESS 1 MORE

(say "Seventy")

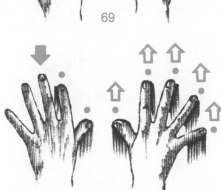

YOU HAVE EXCHANGED 69 FOR 70

Count to 79

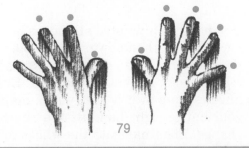

79

(79) to (80)

PRESS 1 MORE

(say "Eighty")

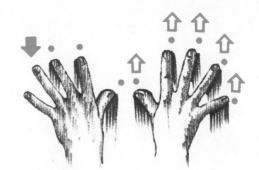

YOU HAVE EXCHANGED 79 FOR 80

Count to 89

89

(89) to (90)

PRESS 1 MORE

(say "Ninety")

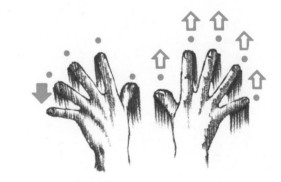

YOU HAVE EXCHANGED 89 FOR 90

You have manipulated all the exchanges in the 9s to 10s category. Run a quick marathon now from 1 to 99, using unit-by-unit, finger-by-finger progression. Try to maintain a steady pace with your oral counting, forcing your fingers to keep up. Every time you reach an exchange hurdle, jump over it with gusto. Slap down your 10s finger and flip up your entire right hand with a rhythm that is unbroken at every exchange. Then navigate the whole course again with your eyes closed.

72

Try doing this to music. Then you won't be able to cheat by slowing down at the hurdles. When you can run from start to finish without tripping or missing a beat, then you're ready to move onto a new track.

Congratulations!

CORRECT COMBINATIONS (They Open All Doors)

Every one of your fingers has gotten into the act. You're able to count by units to 99 with confidence, executing exchanges rhythmically and cleanly. The pattern and regularity of these manipulations are becoming more and more comfortable and familiar. Only one final skill remains for you to perform automatically and you'll be ready to tackle computations—which are the ultimate goal in Fingermath.

You learned earlier to press single-digit numbers instantly on command, using only your right hand. Now you have to develop just as much skill in combining both hands to press double digits. Do you remember how I stressed the distinction between recognizing a two-hand press as a *whole number* rather than the sum of one number plus another number? I am reminding you of that because, when you tackle actual calculations, you'll have to recognize instantly combinations of fingers on both hands as whole numbers.

Go through all the exercises on page 75. Then ask your helper to call them out to you for another go-around. Make a habit of repeating this drill every day.

PRESS 19   PRESS 6

   CLEAR your fingers   CLEAR

PRESS 32   PRESS 49

   CLEAR your fingers   CLEAR

PRESS 50   PRESS 22

   CLEAR   CLEAR

PRESS 13   PRESS 71

   CLEAR   CLEAR

PRESS 87   PRESS 4

   CLEAR   CLEAR

PRESS 15	PRESS 13
CLEAR	CLEAR
PRESS 7	PRESS 60
CLEAR	CLEAR
PRESS 12	PRESS 68
CLEAR	CLEAR
PRESS 80	PRESS 18
CLEAR	CLEAR
PRESS 8	PRESS 29
CLEAR	CLEAR
PRESS 20	PRESS 33
CLEAR	CLEAR
PRESS 23	PRESS 5
CLEAR	CLEAR
PRESS 97	PRESS 9
CLEAR	CLEAR
PRESS 6	PRESS 70
CLEAR	CLEAR
PRESS 75	PRESS 14
CLEAR	CLEAR

Practice is not a one-day affair. The exercises should be repeated every day for at least five minutes. When you've had your fill of oral commands, make up your own written list of double-digit numbers (not in sequence, please) and work with it for a change. Get accustomed to seeing a number and sensing which finger combinations will make it happen. Learning these combinations and pressing them confidently are among the most important Fingermath skills you can master, ones that open the doors to every arithmetical operation.

Double Digit Combinations

Instructions: First press what you see and observe your fingers. Then call out the number.

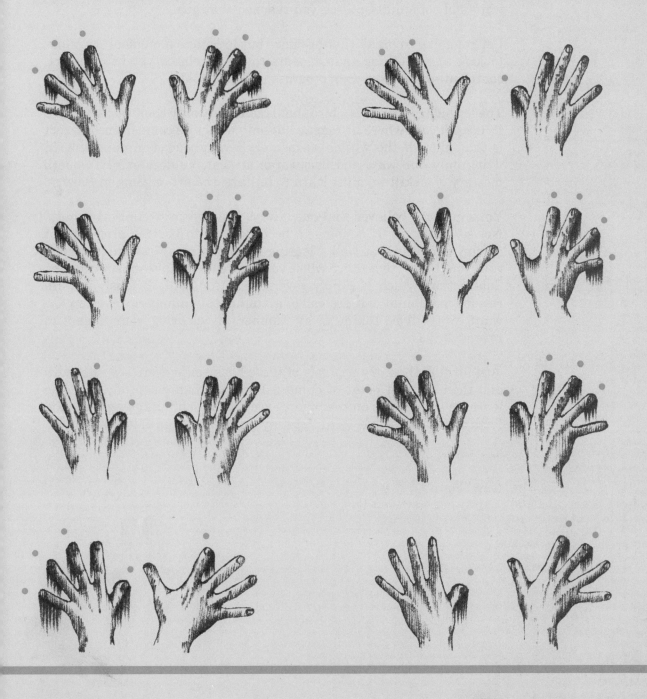

COUNTING IN REVERSE

You've reached a major juncture. You have learned to recognize and press all the numbers up to 99. And I promised to throw you into the waters of computation once you reached that goal.

I'm not going to break that promise, but let's take a momentary pause to look at something that appears to be unrelated yet will, in fact, contribute greatly to your progress in Fingermath.

The legendary whiffle is a bird that is said to fly only backwards because he doesn't care where he's going but only where he came from. The fact is that, though the whiffle seems very wrongheaded in persisting in flying only one way, and backwards at that, he does exhibit superb mastery of a skill we must learn to imitate: the art of going in reverse.

You may have believed that one always moves forward in Fingermath. Not so. Later on, it is going to be necessary to become proficient at moving backwards as well. Clearing numbers in reverse sequence is a skill you'll have to acquire when you are ready to take on subtraction. Subtraction, which is simply reverse addition, will obviously require reverse manipulations. Instead of accumulating numbers, pressing forward, you will be taking away numbers by clearing your fingers in reverse order.

Just for fun, then, try a couple of these backwards exercises right now and then make a point of doing a few like them each day. It's like learning your scales on the piano, except that you'll be going back and forth instead of up and down. Anyway, you've got to go in two directions.

PRESS 4

(say "Four")

Now CLEAR 1

(say "One")

Note carefully that you are clearing the pinky first and that you are beginning your clearing count with "One."

Whenever you wish to clear units, you must begin with the *latest* finger pressed. The rule then is

LAST FINGER IN IS FIRST FINGER OUT

Clear pinky first, then ring finger, then middle, then index. We'll deal with your thumbs later on.

CLEARING SEQUENCE (Last In, First Out)

Pinky Ring Middle Index   Index Middle Ring Pinky

LEFT   RIGHT

PRESS 3

(say "Three")

Then CLEAR 1

(say "One")

Read Your Fingers . . . (2) Then Clear Your
Fingers

PRESS 40

(say "Forty")

Then CLEAR 10

(say "Ten")

Read your Fingers . . . (30) Then Clear Your
Fingers

Observe again that you *say the amount being cleared.* Then what you
read on your fingers gives you the result.

THUMBS UP (It's Still "Last In, First Out")

Here's where the real fun begins! When your thumbs get into the act, continue to think in reverse. Just as you did a front flip to go up from 4 to 5, now you will do a back flip to go down from 5 to 4. Watch!

START

WITH

5

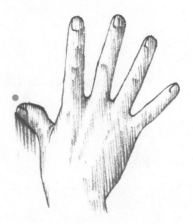

TO CLEAR 1

Do a Back Flip

SIMULTANEOUSLY:

CLEAR 5 and

PRESS 4

(say "One")

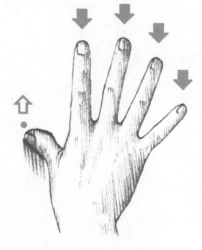

Now try a few back flips, going from 5 to 4. Then go forward from 4 to 5. If you practice the front and back flips interchangeably, neither one will throw you! In all cases, begin with a count of "One," whether you are going forward or in reverse.

Now ask your helper to call out the following commands. Be certain you Clear only *one finger at a time,* saying aloud *how much you are clearing.*

PRESS 4

Clear 1
Read Your Fingers ("Three")
Clear Your Fingers

PRESS 3

Clear 2
Read Your Fingers ("One")
Clear Your Fingers

PRESS 2

Clear 2
Read Your Fingers ("Zero")
Clear Your Fingers

PRESS 4

Clear 3
Read Your Fingers ("One")
Clear Your Fingers

PRESS 30

Clear 20
Read Your Fingers ("Ten")
Clear Your Fingers

PRESS 40

Clear 30
Read Your Fingers ("Ten")
Clear Your Fingers

PRESS 20

Clear 10
Read Your Fingers ("Ten")
Clear Your Fingers

PRESS 30

Clear 20
Read Your Fingers ("Ten")
Clear Your Fingers

PRESS 40

Clear 20
Clear 10 more
Read Your Fingers ("Ten")
Clear Your Fingers

PRESS 4

Clear 1
Clear 2 more
Clear 1 more
Read Your Fingers ("Zero")

PRESS 3

Clear 1
Clear 2 more
Read Your Fingers ("Zero")
Clear Your Fingers

Here's a final exercise to keep you in shape each day while you'll be concentrating mainly on forward actions.

REVERSE COUNT FROM 10 TO 0

(10) **PRESS 10**

(say "Ten)

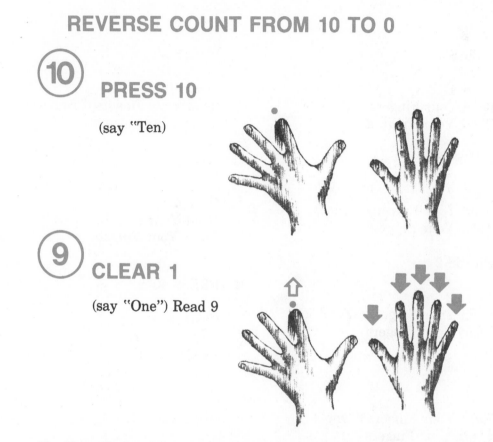

(9) **CLEAR 1**

(say "One") Read 9

You'll really have to concentrate on this exchange. There's still a devil lurking here, the same as you encountered when you were moving in a forward mode from 9 to 10. Once again, if you practice in both directions interchangeably, going forward and then backwards, and always calling out the exchange, both manipulations will soon become automatic. If you have not been saying all your manipulations aloud, you are costing yourself valuable time in becoming truly skilled at Fingermath. By calling out every step, you are adding another sensory aid to those already in use. You can feel your fingers pressing, you can see them pressing, and by calling out you can hear every number. This total sensory combination works to your advantage. I only wish there were some way that you could taste and smell every manipulation.

But back to reverse counting! You should still be pressing 9.

(4) CLEAR 1 MORE

(say "One") Read 4

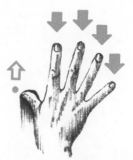

(5 to 4 exchange)

(3) CLEAR 1 MORE

(say "One") Read 3

(2) CLEAR 1 MORE

(say "One") Read 2

(1) CLEAR 1 MORE

(say "One") Read 1

Now

8 CLEAR 1 MORE

(say "One") Read 8

7 CLEAR 1 MORE

(say "One") Read 7

6 CLEAR 1 MORE

(say "One") Read 6

5 CLEAR 1 MORE

(say "One") Read 5

(0) CLEAR 1 MORE

(say "One") Read 0

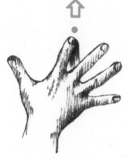

Do the whole sequence from 10 to 0 again and try to think it through without diagrams. As soon as you grasp the pattern, do it with your eyes closed. Always count aloud, calling out the amount you clear. Finally, go forward from 1 to 10, then backwards from 10 to 0. Practice each sequence alternately. Try to press or clear your fingers at an even pace. Be certain you do not falter at the critical points of exchange; when going from 4 to 5, 9 to 10, 5 to 4 and 10 to 9. Here, again, music can help you to establish a steady rhythm for each manipulation. Take plenty of time with these exercises. They're important. Be certain to clear only one finger at a time, counting aloud when each finger is cleared.

Basic Addition and Subtraction

ADDITION
The Heart and Soul of Fingermath

Addition is an operation that most of us have done on our fingers in some form or other since childhood. The big plus with Fingermath is that your fingers will now be used scientifically. There's no magic to it. In fact, with minor verbal modifications, addition will employ the identical progressions, sequencing, and manipulations that you have been using for straightforward Fingermath counting.

Every one of the skills you've learned so far is going to serve you well when you perform calculations of addition or, indeed, any of the other three operations of arithmetic as well. You'll soon be handling all four with ease—addition, subtraction, division, and multiplication. Remember that old chestnut, "Be the first on your block to. . . ." Well, before you're finished you may well be the first on your block to do many truly unusual and remarkable computations, all without any electronic gadgets. You'll be turning upside down the tired old boast of "Look, no hands!" Your ten-finger calculator will always be at hand and it will never run out of batteries.

So flex those digits and get ready to put all your hard-earned skills to work.

For starters, I'm going to ask you to stay with unit-by-unit counting. Before long, you'll have the hang of it and be able to move swiftly onto more advanced levels, taking advantage of every shortcut you've already used—and then some. Once again, it's only a matter of practice.

You might as well take advantage of any time not otherwise occupied, and run through your Fingermath manipulations. While you're driving your car, you can easily keep both hands on the wheel and your eyes on the road, and still practice pressing and adding up numbers. If you've got things on your mind when you go to bed, try lying on your back with your arms relaxed straight out, palms down, and your eyes closed. You'll press your way to sleep, and it beats counting sheep. You're excused in this one instance from counting aloud! Or when you're taking it easy in your favorite armchair, listening to the stereo, your hands are perfectly set for Fingermath. Just ten minutes a day will steadily improve your accuracy and speed. What's more, you'll find that short practice periods daily are better than trying to cram all your exercises into a single long workout.

Here is a daily exercise program that might fit comfortably into those ten minutes. Naturally, if there are any areas where the going seems especially rough, you should spend the most time there. The following is a good general program for all areas.

Scales

Count forward by units from 1 to 99. Then count backwards by units from 10 to 0. In both directions you should aim at getting the progression to feel comfortable and automatic. It will help if you say the number you are pressing. For example, go forward saying one, two. three, four, etc.... and go backwards to the ten, nine, eight, seven, etc....

Recognition of Finger Values

Ask your helper to call out single- and double-digit numbers at random while you press them instantly on command. Quicken the pace gradually. Work the entire range between 1 and 99.

Exchanges

Do the series of bridges from the 9's to the 10's, calling out the numbers being pressed, such as twenty-nine,–thirty; nine, ten; forty-nine, fifty.

Now it's time you began using a pencil. While you'll be handling some examples orally, you will also start to solve larger, more complex problems that involve writing. At first, you may be inclined to pick up your pencil only when you need to write down a number. Later, as your fingers become accustomed to holding it, you'll want to keep the pencil cradled in your fingers all the while as you do your Fingermath manipulations. Your hand will look like this:

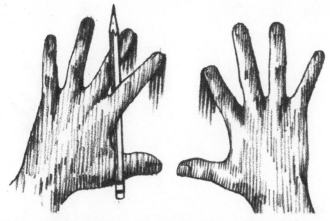

Keeping the pencil in hand during manipulations improves your speed and keeps your fingers in place.

Not having to stop to pick up and put down the pencil every time it's needed, your fingers will remain in an active position without losing speed. You'll find that holding a pencil throughout even one practice session is enough to make you feel at home with one thereafter. (Personally, I find pens too fat for comfort. And if you're working with a small child, look through your desk for a short skinny pencil that can fit comfortably in tiny fingers. Otherwise, a standard pencil will do.)

You will recall that when you were learning to accumulate numbers, to understand sequence, and recognize finger values, my instructions always read; "Press 1 more–and 1 more–and 1 more. . . ." For addition, you'll want to handle larger numbers, so the instruction "Press 1 more" is of extremely limited value. There's an easy way to meet that objection. Numbers to be added to one another (addends) are a snap even when they are larger numbers. I'll demonstrate by starting small and working with you up to the top.

$$1 + 1 = \square \quad \text{(How Many)}$$

First,

PRESS 1

(say "One")

Plus 1 more (say "One")

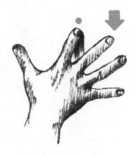

Now read your fingers.

How many?

Correct. Two

CLEAR YOUR FINGERS

$$1 + 2 =$$

First,

PRESS 1

(say "One")

Plus 2 more

STOP

Here's your first verbal departure from a simple "1 more."

When adding any number to another, always begin your count for the new number with "one." That's a rule you must never break if you want to end up with the correct result.

Now continue. You were at 1 and I had just given a new instruction:

Plus 2 more
(say "One–Two")

Now read your fingers.

How many?

Correct. Three

CLEAR YOUR FINGERS

$$1 + 3 =$$

PRESS 1
(say "One")

Plus 3 more
(say "One

"Two

"Three")

Now read your fingers.

How many?

CLEAR YOUR FINGERS

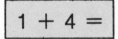

PRESS 1
(say "One")

Plus 4 more
(say: "One

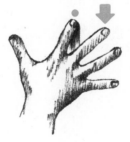

"Two

"Three

"Four")

(Exchange)

Read your fingers.

How many?

Note that although you say "Four," your fingers have registered 5. It is important to disregard the values registered by your fingers *during* the addition until a final result is to be read. This will become even more evident when you get to adding several numbers.

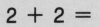

Add Numbers, Counting Aloud. When Finished, Your Fingers Should Be Pressed As Shown On The Right.

(Totals To 4)

$2 + 1 = 3$

$1 + 3 =$

$2 + 2 =$

$1 + 2 =$

$3 + 1 =$

$1 + 1 =$

Repeat each example having someone call out the numbers. As each is completed, the instruction "Clear your fingers" should be given.

$1 + 1 + 1 =$

$1 + 2 + 1 =$

$2 + 1 + 1 =$

$1 + 1 + 2 =$

$2 + 2 + 0 =$

$1 + 3 + 0 =$

Repeat each example having someone call out the numbers. As each is completed, the instruction "Clear your fingers" should be given.

BASIC ADDITION AND SUBTRACTION

$1 + 2 + 2 =$

$2 + 1 + 2 =$

$1 + 1 + 2 + 1 =$

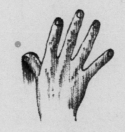

$$2 + 2 + 1 =$$

$$4 + 1 =$$

$$1 + 4 =$$

Repeat each example having someone call out the numbers. As each is completed, the instruction "Clear your fingers" should be given.

BASIC ADDITION AND SUBTRACTION

$$\boxed{1 + 5 =}$$

PRESS 1
(say "One")

Plus 5 more
(say "One")

"Two"

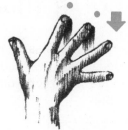

"Three"

"Four"

"Five"

Read your fingers.

How many?

EXCHANGE

CLEAR YOUR FINGERS

Let's replay this same example and apply some of the more advanced knowledge you learned earlier. Think now of the intermediate level technique in which you employ the direct press of 5 if your thumb is available:

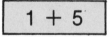

$$1 + 5$$

PRESS 1
(say "One")

Plus 5 more
(say "Five")

Read your fingers.

How many?

CLEAR YOUR FINGERS

Just a word of advice. If you're working with a child, hold off for a time before using this shortcut. It's important not only to come up with answers but to understand how the answers are reached. By repeated unit-by-unit counting, a child learns the route to each answer. Later, when you're sure the route is completely familiar, you can move your youngster up to this next level.

As for yourself, you will be making only limited use of the direct press of 5 for the time being. Obviously, when your thumb has already been pressed and hence is not available, you'll have to rely on a unit-by-unit count. Shortly, I'll be taking you through a new Press-and-Clear manipulation that bypasses such basic-level techniques. For the moment, let's continue.

PRESS 1

(say "One")

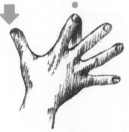

Plus 6 more

(say "Five")

"Six"

Read your fingers.

How many?

CLEAR YOUR FINGERS

That was the first instance where you may *seem* to have jumped over three unpressed fingers—a prohibited move, as you well know. But it's easy to see that nothing illegal took place. The rules say that it's only the units fingers that must be pressed in unvarying sequence. You are always permitted to use the thumb if it's available for a direct press of 5. That's what was done, and you reached 6 by pressing the middle finger, the first units finger free at that time.

Just for practice (and for a child), do this same example adding the 6 with unit-by-unit counting. After pressing the first 1, be sure to begin your new 6 count with another "One." It should look like this:

1 + 6 (by Units)

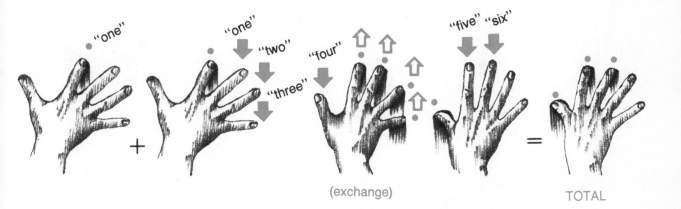

(exchange)   TOTAL

It's time you started working with numbers larger than 1. Since you can instantly press any number to 99, get into the habit of doing that with the first number in any example. This can be done by a child as well. Try the following for starters, registering the first number with an instant press and then adding the second number unit by unit. Finally, use a direct 5 press where possible.

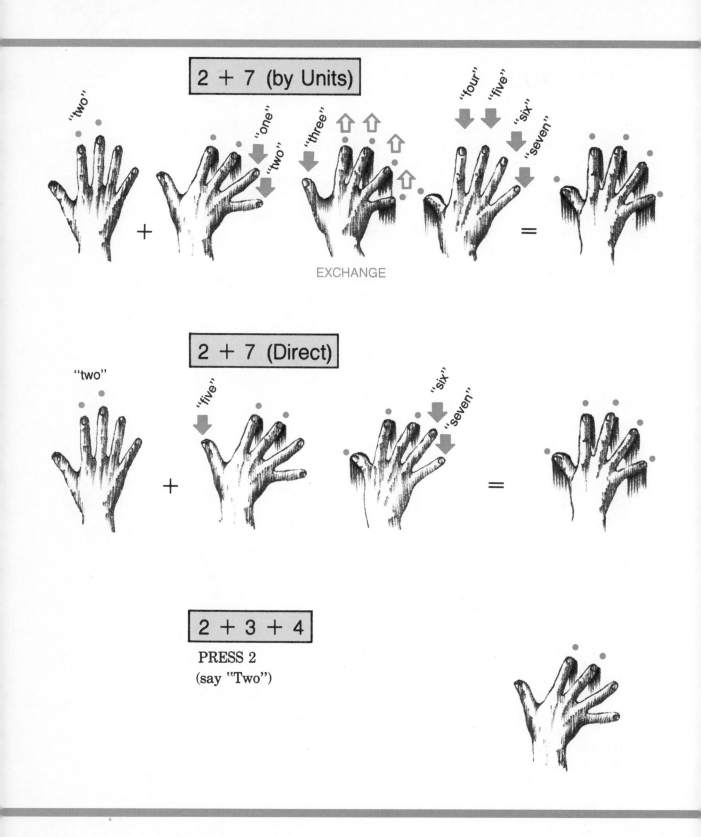

2 + 7 (by Units)

"two"   "one" "two"   "three"   "four" "five" "six" "seven"   =

EXCHANGE

2 + 7 (Direct)

"two"   "five"   "six" "seven"   =

2 + 3 + 4

PRESS 2
(say "Two")

108

PLUS 3 More
(say "One")

"Two"

"Three"

EXCHANGE

PLUS 4 MORE
(say "One")

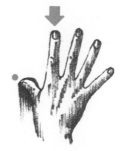

"Two"

"Three"

"Four"

Read your fingers.

How many?

CLEAR YOUR FINGERS

Add Numbers, Counting Aloud. When Finished, Your Fingers Should Be Pressed As Shown On The Right.

(Totals To 9)

$3 + 2 =$

$2 + 2 + 2 =$

$1 + 4 + 1 =$

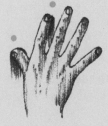

$$2 + 3 + 4 =$$

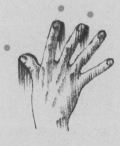

$$3 + 2 + 3 =$$

$$1 + 2 + 2 + 2 =$$

Repeat each example having someone call out the numbers. As each is completed, the instruction "Clear your Fingers" should be given.

Add Numbers, Counting Aloud. When Finished, Your Fingers Should Be Pressed As Shown On The Right.

(Totals To 9)

$2 + 3 + 3 =$

$1 + 3 + 2 + 2 =$

$2 + 5 + 1 + 1 =$

$$3 + 1 + 3 =$$

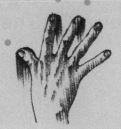

$$1 + 2 + 3 + 1 =$$

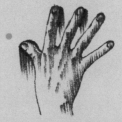

$$4 + 1 + 3 =$$

Repeat each example having someone call out the numbers. As each is completed, the instruction "Clear Your Fingers" should be given.

Your left hand has rested long enough. Now you can put it to work and see how beautifully both hands can add on and store numbers and present you with a simple total. You will recognize the total instantly because of all the time you have spent developing the skill of quickly sensing finger values.

$$3 + 2 + 5 =$$

PRESS 3

PLUS 2 MORE

(say "One")

"Two"

(exchange)

PLUS 5 MORE
(say "One")

"Two"

"Three"

"Four"

"Five"

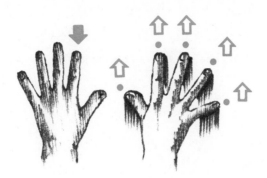

Read your fingers.

How many? (Enter total)

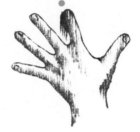

CLEAR YOUR FINGERS

If you're coming out with the wrong totals, my guess is that you are failing to begin the counting for each new added number (addend) with a spoken "One." Remember, ignore your finger values as you are accumulating each new addend or you'll get totally confused. To illustrate, look at the last example. When you started adding the 2, your fingers read 4, and when you started adding the 5, your fingers read 6. Pay no attention to these subtotals but concentrate on talking yourself through the proper finger sequence. When you've finished with the last addend, *then* read your fingers for the total.

Are you trying to keep your pencil in your hand while you do these manipulations?

Add Numbers, Counting Aloud. When Finished, Your Fingers Should Be Pressed As Shown On The Right.

(Totals To 10)

$$2 + 3 + 5 = 10$$

$$4 + 2 + 3 + 1 =$$

$$3 + 1 + 4 + 2 =$$

118

$6 + 1 + 2 + 1 =$

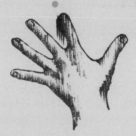

$5 + 2 + 3 =$

$4 + 1 + 4 + 1 =$

Repeat each example having someone call out the numbers. As each is completed, the instruction "Clear your Fingers" should be given.

4 + 5 + 7 (by Units)

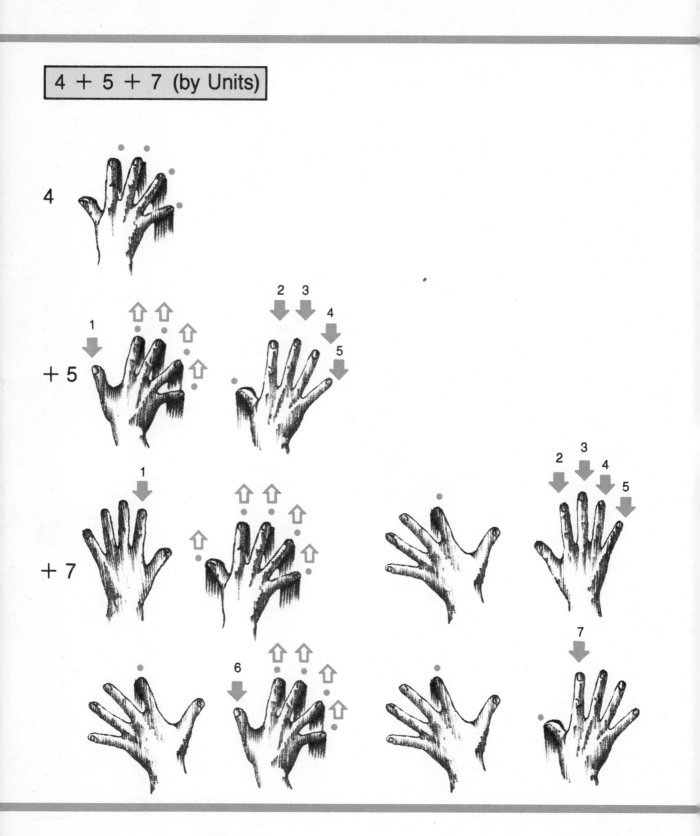

4 + 5 + 7 (Direct)

In this direct method, you must be very careful, when pressing 5, that you do not clear an already established 4. You are not exchanging 4 for 5, a manipulation that adds only 1. Rather, you are adding 5 more to the 4. There's a big difference. Watch out for this devil; he wears many disguises!

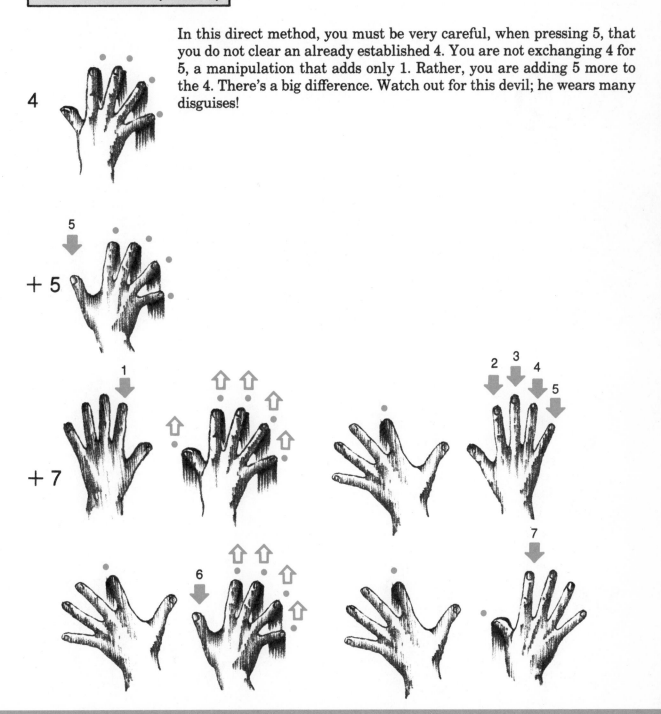

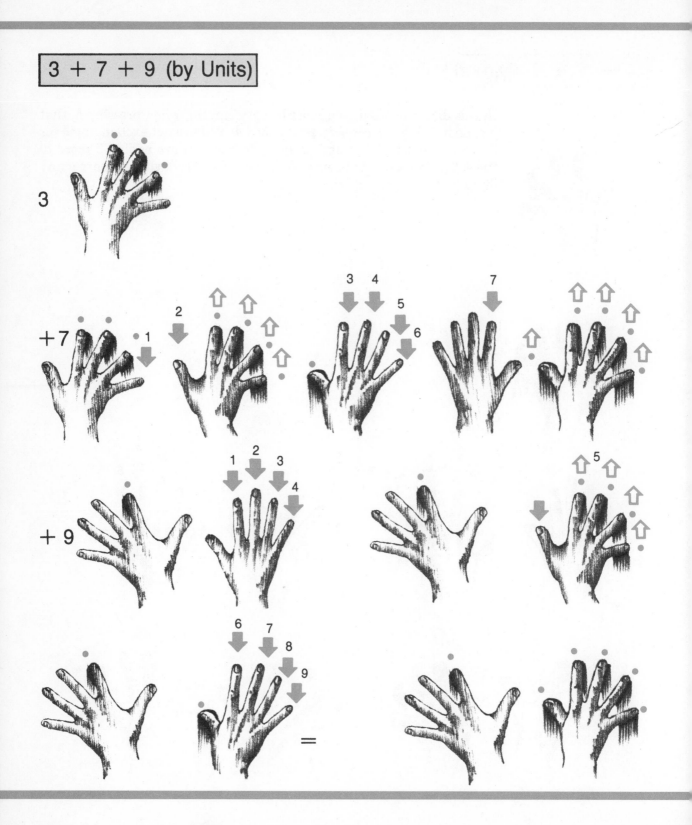

122

3 + 7 + 9 (Direct)

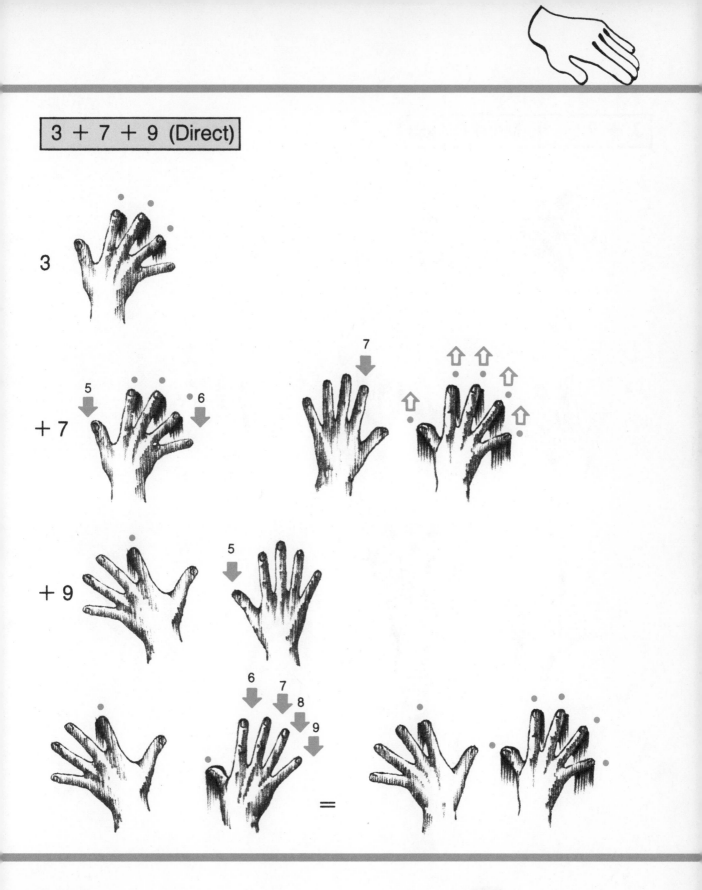

3 + 7 + 9 (More Direct)

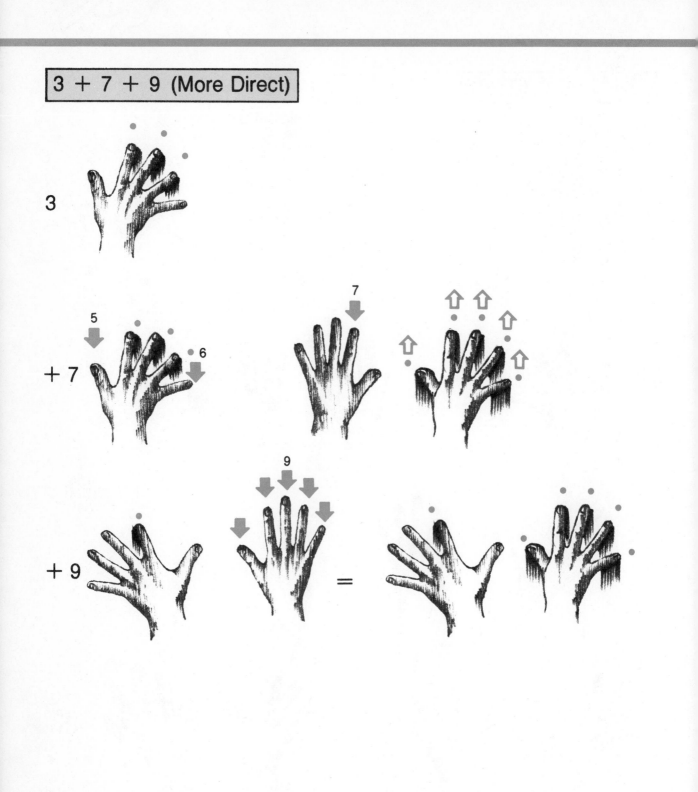

This last set reveals another shortcut. Having added 7, you found your entire right hand was clear. Whenever this happens, you can go to a direct press of any single-digit whole number. No matter which whole single-digit number may come up when your right hand is clear, treat it as you would when *starting* an example. That is, press it as a whole number, skipping any other less direct count. This approach is fine for children also, provided that they have learned to press whole numbers.

As if this weren't direct enough, there remains a still more direct way of calculating this or any other example. It is so direct, in fact, that even a child would be able to add up accurately and quickly those three numbers in the last examples. So hang in there, don't skip any work pages, practice your "scales" each day, review instant recognition of single and double digits, say your numbers aloud, and work with pencil in hand. Practice at times with your eyes closed, and don't try to cover too much ground in any single session.

One thing more. You should be practicing counting in reverse by units *every day.* Get so good at it that you are capable of clearing fingers, one-by-one, from 99 down to 0, with no break in the rhythm as you steadily count aloud. We will deal with subtraction soon, and this accomplishment of clearing numbers will help you to subtract numbers as easily as you have learned to add them.

Add Numbers, Counting Aloud. When Finished, Your Fingers Should Be Pressed As Shown On The Right.

(Totals To 14)

$$4 + 3 + 4 =$$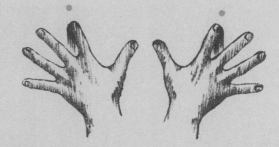

5 + 4 + 1 + 4 =

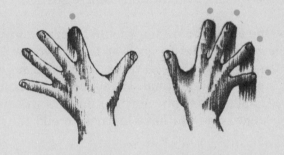

6 + 5 + 3 =

4 + 2 + 5 =

BASIC ADDITION AND SUBTRACTION

$7 + 4 + 2 =$

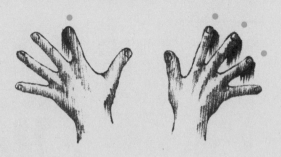

$8 + 2 + 4 =$

Repeat each example having someone call out the numbers. As each is completed, the instruction "Clear your Fingers" should be given.

It is a simple matter now to continue in a similar pattern that is slightly more advanced, in that it begins with a direct press of a whole double-digit number. As before, this gives you the advantage of a fast start with the first number in any example. (If you—or your child—are not yet confident enough to execute the direct press of a whole number, more practice is needed. Call out numbers for one another, at random and at a good, steady pace. Practice until you respond automatically to commands for any numbers from 1 to 99.)

$$\boxed{14 + 3}$$

(THIS IS TRICKY . . . WATCH IT!)

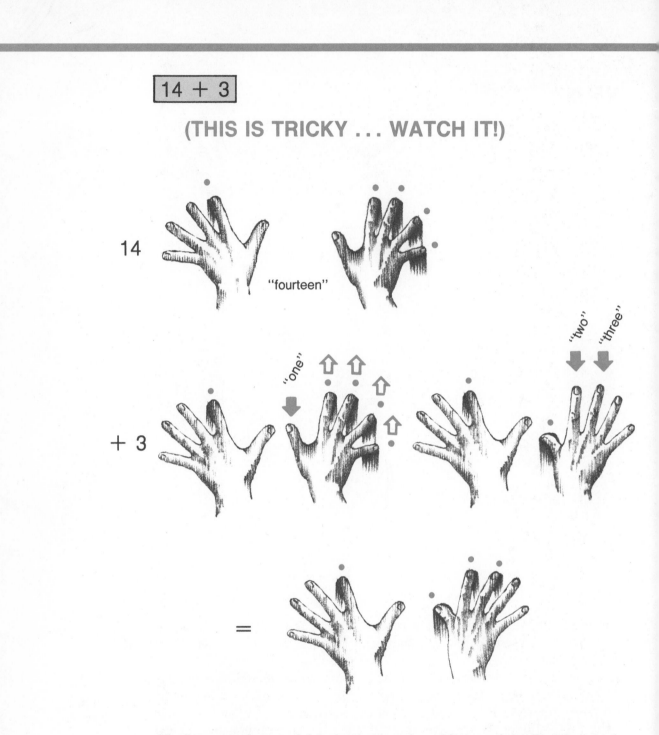

14 "fourteen"

+ 3 "one" "two" "three"

=

You've got to be careful in beginning the addition of the 3 with a 4 to 5 exchange. You're pressing only one unit at a time. If you don't execute that front flip, you'll end up 4 units overweight. Daily practice of unit-by-unit counting to 99 protects you from falling into traps like this.

BASIC ADDITION AND SUBTRACTION

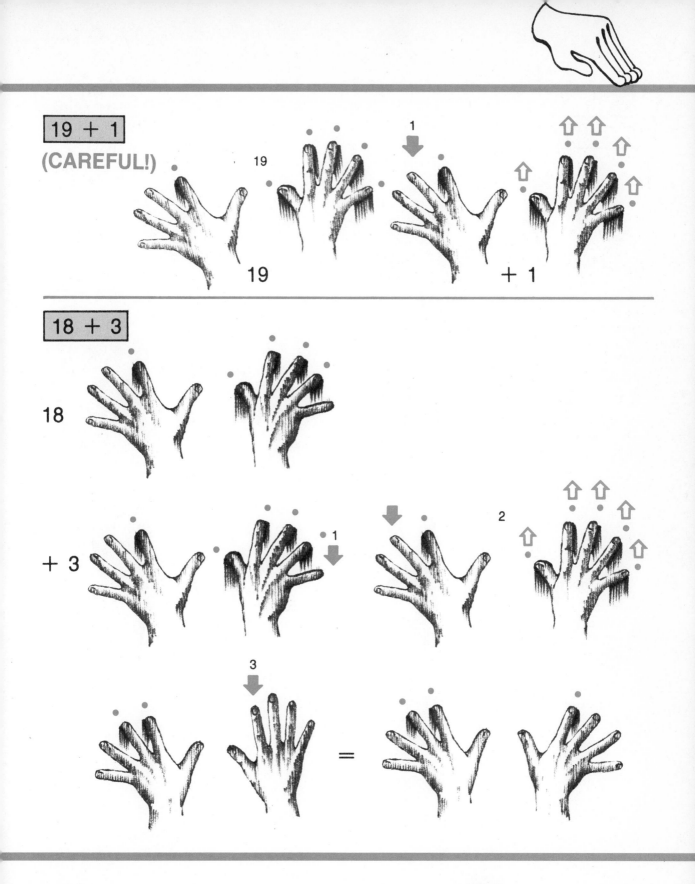

19 + 1
(CAREFUL!)

19

19

+ 1

18 + 3

18

+ 3

1

2

3

=

129

19 + 6

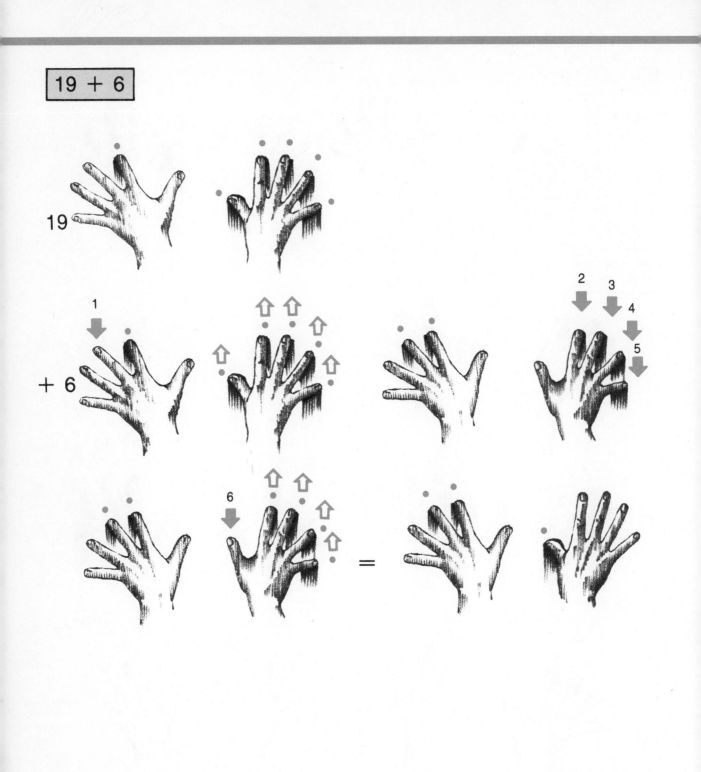

BASIC ADDITION AND SUBTRACTION

23 + 7 (by Units)

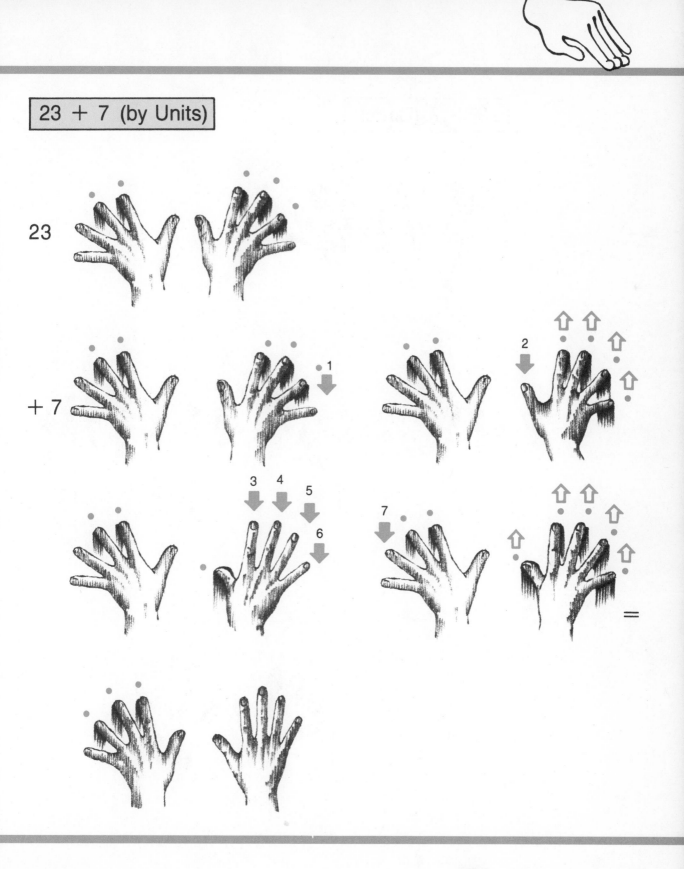

23 + 7 (Direct)

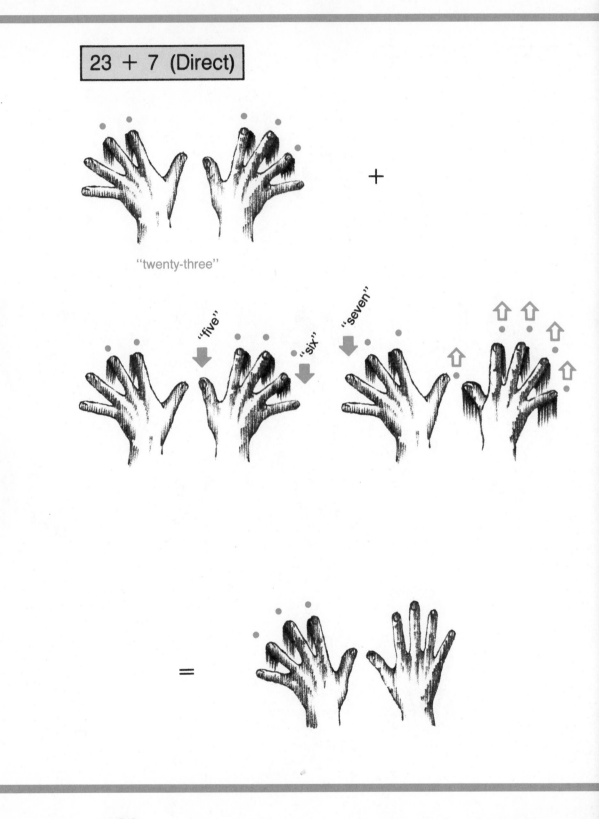

"twenty-three"

+

"five" "six" "seven"

=

BASIC ADDITION AND SUBTRACTION

Go back now and go through the section on addition from the beginning. Ask someone to read out the addends of each example so that you'll get some practice handling numbers sight unseen. Do *all* of them, right through these latest ones with double digits. Then ask your helper to make up some new examples, deliberately trying to set traps for you, which you, of course, must try your best to avoid.

Adding Double Digits. Add Numbers, Counting Aloud. When Finished, Your Fingers Should Be Pressed As Shown On The Right.

$12 + 9 =$

$21 + 6 + 8 =$

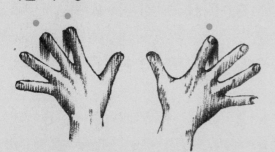

$9 + 1 + 14 =$

$37 + 4 + 7 =$

Repeat each example having someone call out the numbers. As each is completed, the instruction "Clear your Fingers" should be given.

Calculate Both Ways . . . Horizontally and Vertically
Addition

$8 + 2 + 5 + 7 + 3 + 9 + 8 + 4 + 8 + 5 + 9 =$

$6 + 5 + 8 + 2 + 9 + 6 + 5 + 3 + 5 + 9 + 6 =$

$2 + 4 + 6 + 3 + 6 + 7 + 5 + 9 + 8 + 6 + 3 =$

$3 + 6 + 8 + 7 + 3 + 5 + 9 + 8 + 9 + 8 + 1 =$

$7 + 2 + 6 + 3 + 7 + 9 + 5 + 7 + 8 + 6 + 9 =$

$2 + 1 + 5 + 6 + 9 + 8 + 3 + 4 + 7 + 9 + 6 =$

$5 + 7 + 8 + 9 + 2 + 4 + 7 + 8 + 9 + 5 + 7 =$

$1 + 9 + 7 + 3 + 4 + 8 + 6 + 9 + 3 + 4 + 9 =$

$9 + 6 + 8 + 8 + 7 + 7 + 5 + 5 + 3 + 3 + 9 =$

$3 + 6 + 8 + 9 + 5 + 5 + 5 + 6 + 6 + 3 + 3 =$

$8 + 9 + 4 + 5 + 7 + 7 + 3 + 4 + 4 + 8 + 6 =$

Do this page orally also.

Addition In Reverse

Earlier (page 78), I touched on subtraction, remarking that in Fingermath it was done by reversing the finger manipulations used for addition; it requires the same skills turned around. Numbers must be taken away. You must reduce your total by clearing fingers.

This reverse procedure would have been confusing if we had tried to develop it before you were thoroughly familiar with the finger actions of addition. It's like learning to drive a car. Much practice must be given to going forward before you attempt to go in reverse. Otherwise you might go in a direction you couldn't control. By first getting hours of practice with addition, you are not likely to be confused now by a reversal of direction. But take care!

The same devils that cause problems with addition show up again in subtraction. There are only a couple of them, so it's wise to go past them *every day,* observing their peculiarities. This is how to proceed:

Count down from 5 to 4. Count down from 10 to 9. Then count forward and backwards, adding on and taking away values as you make the exchanges between fingers and thumb and between right and left hands.

Remember that to subtract, you clear the right-hand fingers from right to left. So, counting from 4 to 0, first clear the pinky, then the ring

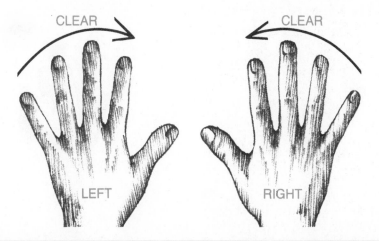

CLEAR   CLEAR

LEFT   RIGHT

finger, then the middle, then the index finger—just the reverse of the sequence for addition. On the left hand, clear from left to right—again the reverse of addition.

Get to recognize the similarities and the differences. With enough practice, your fingers will know which way to move, automatically. Start at various numbers below 10 and count in reverse, unit by unit, to 0. When you feel secure reviewing this simple reverse counting, go on to do the workpages in simple subtraction. Here's how to proceed.

Say each example aloud. Press the top number (the minuend), saying it at the same time. Then subtract the bottom number (the subtrahend) by counting aloud in the normal sequence from one on, at the same time continuing to clear units until the subtrahend is counted out. Then read your fingers; the value they register is the answer (the difference).

For example:

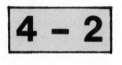

PRESS 4

CLEAR (subtract) 2, Unit-by-Unit, counting aloud from "One".

"One"

"Two"

Now, read your fingers.

The answer (difference) is 2.

Here's another:

$$5 - 3$$

PRESS 5

CLEAR (subtract) 3, Unit-by-Unit, counting aloud from "One".

"One"

"Two"

"Three"

NOW, READ YOUR FINGERS.

The answer (difference) is 2.

Try this example which requires both hands!

12 – 4

PRESS 12

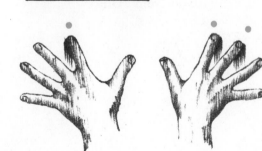

Clear (subtract) 4 Unit-by-Unit, counting aloud from "One".

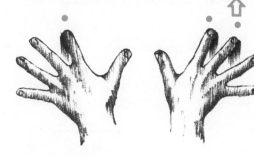

"One"

"Two"

"Three"
(10 to 9 exchange)

"Four"

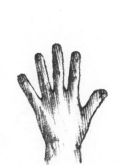

139

READ YOUR FINGERS.

The answer (difference) is 8.

And ... one more, just for good measure, putting two of the reverse exchanges to work:

Press 16

Clear (subtract) 7, counting aloud from "One".

"One"

"Two"

(5 to 4 exchange)

"Three"

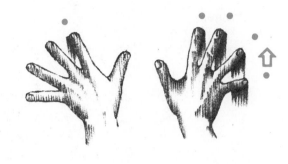

"Four"

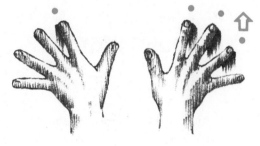

"Five"

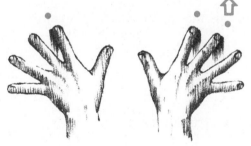

"Six"

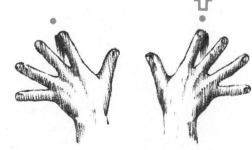

"Seven"

(10 to 9 exchange)

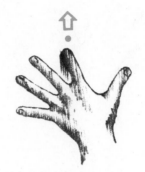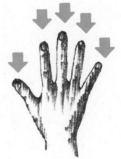

READ YOUR FINGERS.

The answer (difference) is 9.

For the present, subtract only one column at a time. Don't attempt double digit subtraction until further on, when the clearing of double digit numbers ... through practice ... becomes simple. In other words, when faced now with an example like this:

$$\begin{array}{r} 47 \\ -16 \\ \hline \end{array}$$

First subtract the ones (7–6 is 1).

Next subtract the tens (4–1 is 3).

Simple Subtraction

9 −4	8 −3	7 −2	6 −1	5 −4
9 −1	5 −1	4 −3	3 −2	2 −1
1 −1	8 −7	7 −6	9 −8	8 −6
34 −22	57 −32	86 −74	94 −34	63 −11
9 − 6 =	12 − 3 =	22 − 9 =	37 − 8 =	62 − 7 =
8 − 4 =	9 − 1 =	10 − 1 =	6 − 1 =	7 − 2 =
9 −5	8 −1	7 −5	9 −7	6 −3
67 −24	35 −15	84 −73	77 −56	28 −23

Simple Subtraction

8 − 2 − 2 − 2 ————	9 − 5 − 1 ————	9 − 2 − 3 − 1 ————	8 − 4 − 3 ————
19 − 9 = ☐	16 − 8 = ☐	18 − 3 = ☐	18 − 9 = ☐
6 3 − 4 − 5 ————	7 − 6 − 2 ————	9 − 3 − 5 ————	9 − 2 5 − 1 ————
6 − 4 = ☐	9 − 5 = ☐	10 − 1 = ☐	5 − 1 = ☐
863 −421 ————	6 5 4 −7 ————	7 − 5 6 − 4 ————	5 6 9 −8 ————

FROM TENS TO UNITS

The last page of exercises has given you plenty of practice manipulating the reverse exchanges from 5 to 4 and from 10 to 9. Now study the other major reverse manipulations which exchange tens for units.

Ⓞ TO ⑲

PRESS 20

say "Twenty"

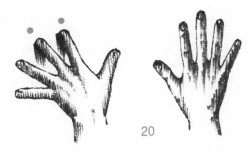

20

Clear 1

say "One"

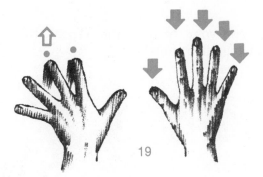

19

Clear your fingers and repeat this exchange five times.

 **30** to **29**

PRESS 30

say "Thirty"

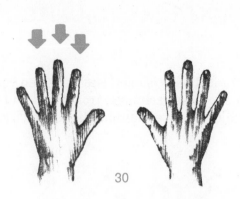

30

Clear 1

say "One"

29

Clear your fingers and repeat this exchange five times.

 **40** to **39**

PRESS 40

say 'Forty"

40

146

Clear 1

say 'One'

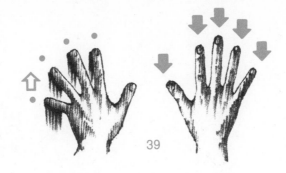

Clear your fingers and repeat this exchange five times.

 **to** 49

This manipulation needs extra practice. Take your time.

PRESS 50

say 'Fifty'

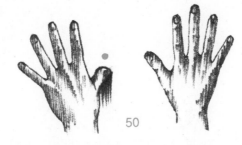

Clear 1

say "One"

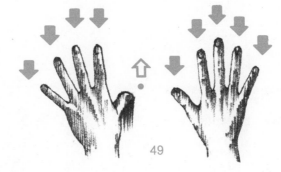

Clear your fingers and repeat this exchange 10 times.

60 TO 59

PRESS 60

say "Sixty"

60

Clear 1

say "One"

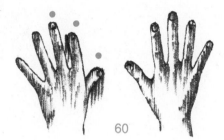

59

Clear your fingers and repeat this exchange five times.

70 TO 69

PRESS 70

say "Seventy"

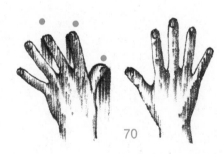

70

Clear 1
say "One"

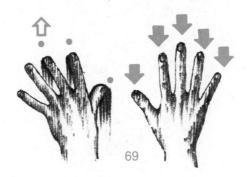

Clear your fingers and repeat this exchange five times.

80 TO 79

PRESS 80
say "Eighty"

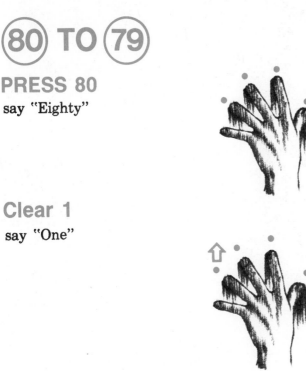

Clear 1
say "One"

Clear your fingers and repeat this exchange five times.

(90) to (89)

PRESS 90

say "Ninety"

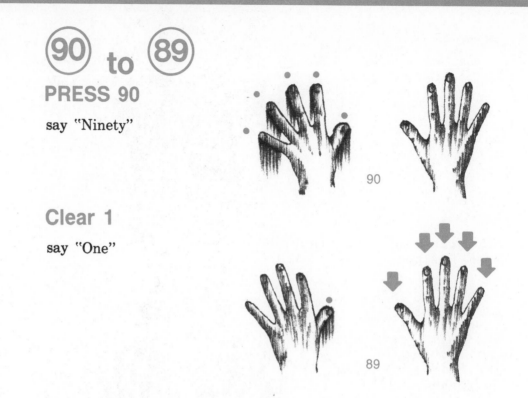

90

Clear 1

say "One"

89

Clear your fingers and repeat this exchange five times.

Every one of these reverse exchanges employs the identical manipulation, except when going from 50 to 49. For extra reinforcement of this maneuver, try alternating several times between the exchange from 50 to 49 and then directly back to 50, like this:

PRESS 50

say "Fifty"

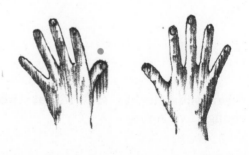

150

PRESS 49

say "Forty-nine"

PRESS 50

say "Fifty"

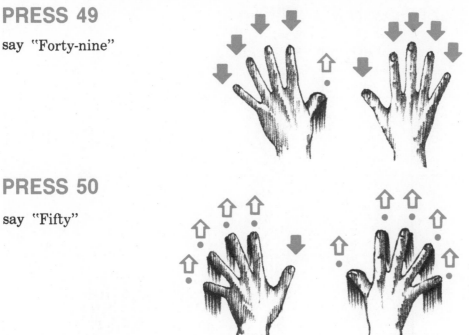

Now you're prepared to take a finger-by-finger, unit-by-unit journey through the entire reverse trail from 99 to 0.

Press 99 and then clear your way slowly and deliberately to zero. Because you have become more confident about reverse manipulations, you can now say the numbers you are pressing (ninety-nine, ninety-eight, ninety-seven, etc.), rather than the amount you are clearing, which would always be the procedure for subtraction.

Repeat this essential exercise until you can perform evenly, never stumbling at the exchanges. When you can move at a deliberate, steady pace, put your skills to music, always counting aloud. Try it with your eyes closed.

This is another exercise that hereafter should be practiced daily.

Now you can dig into the examples that follow. Remember that after pressing the first number you must begin with a new count of "One" for each number that follows, regardless of whether it is to be added or subtracted. Read your fingers for an answer only when the entire calculation has ended.

Subtraction & Addition

After the initial Press, count each number beginning with "One."
This entire page to be repeated orally.

$28 - 9 + 5 = \square$	$87 - 8 + 6 = \square$	$13 + 7 - 9 = \square$
$19 + 9 + 3 - 7 = \square$	$22 + 8 - 1 = \square$	$11 - 2 - 1 - 8 = \square$
$39 + 9 + 5 - 7 = \square$	$77 + 5 - 3 = \square$	$40 - 8 - 6 - 8 = \square$
$77 + 8 - 9 - 7 = \square$	$12 - 9 - 2 + 8 = \square$	$86 - 9 - 9 - 9 = \square$
$59 + 4 - 8 - 5 = \square$	$48 + 8 - 9 = \square$	$39 + 6 + 6 - 9 = \square$
$4 + 9 + 9 - 5 = \square$	$90 + 9 - 8 = \square$	$67 - 8 - 9 = \square$
$56 - 9 - 6 = \square$	$49 + 1 - 2 = \square$	$38 + 8 + 8 - 5 = \square$
$48 + 7 + 7 - 3 = \square$	$8 + 9 + 8 - 6 = \square$	$12 - 3 + 9 - 8 = \square$

Addition & Subtraction

Do this entire page orally.

$7 + 9 + 9 - 8 + 6 - 5 + 8 - 3 - 3 - 1 =$

$4 - 3 + 8 + 9 + 7 - 3 - 3 + 8 - 6 + 2 =$

$9 + 8 + 9 + 6 + 8 + 7 + 9 + 9 + 8 + 9 =$

$8 + 7 - 2 + 5 - 4 - 2 - 6 + 8 + 6 - 3 =$

$5 - 3 + 7 - 4 + 9 - 1 + 8 - 7 - 5 + 6 =$

$3 + 9 + 8 - 2 - 7 + 8 - 4 + 6 - 7 - 4 =$

$6 - 4 + 9 + 8 - 2 + 3 - 5 + 8 + 9 + 9 =$

$9 + 8 - 7 - 6 - 3 + 9 - 6 + 7 - 4 - 5 + 3 =$

$8 - 3 - 2 + 6 + 7 - 8 + 9 - 2 + 9 - 1 =$

$7 - 2 + 9 - 3 - 1 + 6 - 2 + 4 + 5 - 6 =$

$9 + 8 - 6 + 5 + 7 - 2 - 5 - 4 + 8 + 9 =$

$2 + 9 + 8 - 6 - 3 + 7 + 6 - 8 + 4 + 5 =$

$6 - 3 + 5 - 4 + 9 + 6 - 4 + 7 - 8 - 9 =$

$23 + 9 + 9 + 9 - 8 - 8 + 5 + 5 + 6 + 6 =$

$46 - 7 + 7 + 7 + 7 - 3 - 3 + 8 + 8 + 8 =$

$11 + 6 + 6 + 6 + 6 - 7 - 8 - 9 - 5 + 4 =$

$54 - 5 - 4 - 8 - 9 + 7 + 2 + 3 - 6 + 9 =$

$37 - 8 - 8 - 8 - 8 + 5 + 5 + 5 + 8 - 9 =$

SUBTRACTION WITH RENAMING

Up to this point, I've made two-line subtraction easy for you by giving only examples in which each digit of the minuend was larger than each digit appearing below it in the subtrahend. That was just to get you warmed up! Now it's time to confront examples of a kind that often give children and adults a hard time.

When I was in elementary school we learned to do these problems by a method called borrowing. Today, teachers are more likely to refer to the same procedure as renaming or regrouping. No matter what you call it, it's a way of making a small number on the top line of a subtraction example, a minuend, large enough temporarily to permit subtraction of a subtrahend that is greater than the minuend. You probably learned it (and it's still taught) by the method shown below. The procedure is workable, but there is an easier, faster way which we'll get to.

First, let's go through the following subtraction as best we know how already.

$$\begin{array}{r} 35 \\ -9 \\ \hline \end{array}$$

You can't subtract 9 units from 5 units so you must borrow ... sorry, rename the minuend this way:

The 3 is in the tens column. It *means* 3 tens (or 30). We'll take 1 of those tens and give it to the 5. That renames the 35 as 2 tens and 15 units. Now the example looks like this:

$$\begin{array}{r} \overset{2\ 15}{\cancel{35}} \\ -\ 9 \\ \hline \end{array}$$

Now you can subtract 9 from 15, leaving 6. Next you can subtract in the tens place. There are no tens to subtract from the 2 in the tens place; so 2 tens minus 0 tens is 2 tens.

$$\begin{array}{r} \overset{2\ \ 15}{\cancel{35}} \\ -9 \\ \hline 26 \end{array}$$

Thus, the answer is 26.

In a similar way, consider this example:

$$\begin{array}{r} 40 \\ -26 \\ \hline \end{array}$$

We can't subtract 6 units from 0 units. So we will rename the minuend this way:

The 4 is in the tens column. It names 4 tens or 40. We'll take away one of the tens and give it to the 0. That renames 40 as 3 tens and 10 units. Now the example looks like this:

$$\begin{array}{r} \overset{3\ \ 10}{\cancel{40}} \\ -26 \\ \hline \end{array}$$

Now you can subtract 6 from 10 leaving 4. In the tens place, you can subtract 2 from 3 leaving 1.

$$\begin{array}{r} \text{Tens} \quad \text{Ones} \\ 3\ \ 10 \\ \cancel{40} \\ -26 \\ \hline 14 \end{array}$$

The answer is 14.

Fingermath can be used in a very straightforward way to do these subtraction examples, working column by column, after you have renamed the minuend. If you have any problems with subtraction facts or if your child has any problems remembering subtraction facts, Fingermath eliminates them. But you do need to know what to do with the minuend *before* you begin working with subtraction.

Subtraction With Renaming

23 − 5	13 − 4	97 −88	62 −49
97 −89	98 −89	18 − 9	24 − 6
55 −29	66 −47	94 − 8	20 −16
44 −28	81 − 3	48 −39	55 −46
42 −24	22 −19	63 − 7	42 −34

SUBTRACTION WITHOUT RENAMING

As you might guess, Fingermath can be used in other ways to help with subtraction. For instance, the examples on the previous page can be worked without renaming the minuend. You just use the idea of subtraction as "taking away".

To solve any similar example, such as 23 – 9, you can count backwards, as described earlier, from 23. Just press 23, then clear 9, starting your count from "One:"

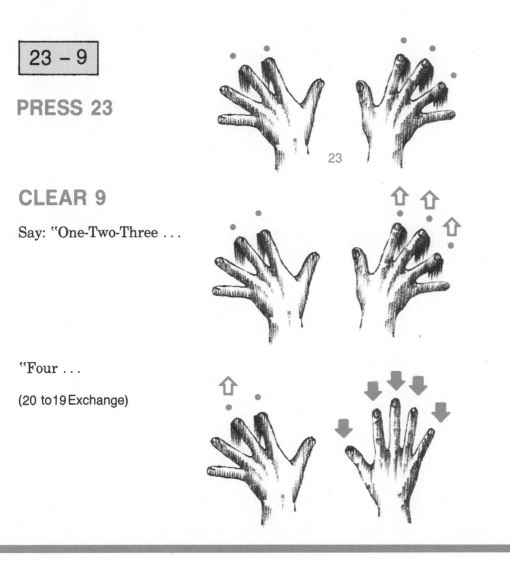

23 – 9

PRESS 23

CLEAR 9

Say: "One-Two-Three . . .

"Four . . .

(20 to 19 Exchange)

"Five, Six, Seven, Eight . . .

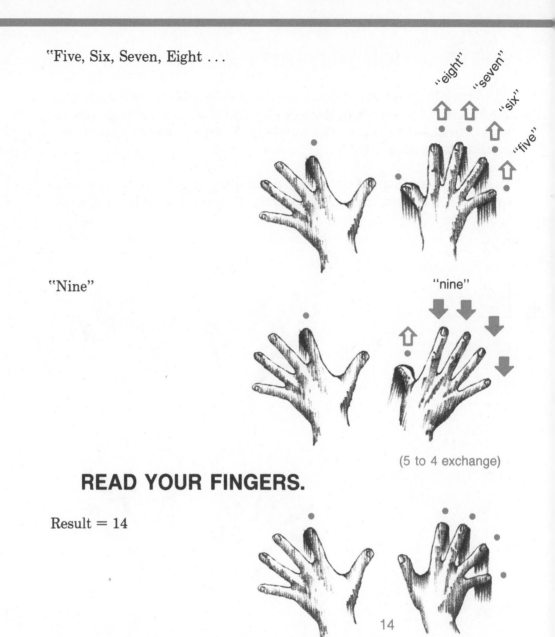

"Nine"   "nine"

(5 to 4 exchange)

READ YOUR FINGERS.

Result = 14

14

As a matter of fact, many people who are good in mental arithmetic solve an example like 40 − 26 by going backwards 20 and then backwards an extra 6. That could be done in Fingermath too.

(If you are working with a youngster at home, however, it's probably best to begin with the units. In that way, you won't be showing the child something that goes against what has been taught in school.)

Using Fingermath to solve this example without renaming:

$$\begin{array}{r} 40 \\ -26 \\ \hline \end{array}$$

Press 40

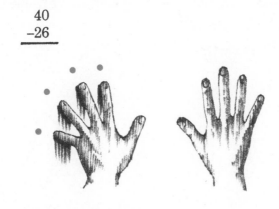

Clear (subtract) 26

This is the first opportunity you have had to use finger-by-finger clearing on the left hand which quickly avoids Unit-by-Unit subtraction at the start of the calculation. Count aloud from Ten.

"Ten"

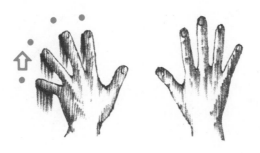

"Twenty"

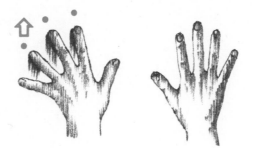

You have cleared 20 but you must clear a total of 26 to complete the example. Now you must continue reverse counting by Units, beginning with "Twenty-One".

"Twenty-One"

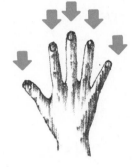

(20 to 19 exchange)

"Twenty-Two"

"Twenty-Three"

"Twenty-Four"

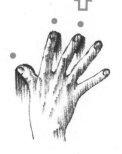

"Twenty-Five"

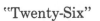

"Twenty-Six"

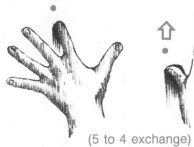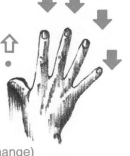

(5 to 4 exchange)

NOW, READ YOUR FINGERS.

The answer (difference) is 14.

It is important to remember that while you are counting the amount being cleared, you must ignore the value shown on your fingers until you have completed subtracting and are ready to read your answer.

Here's another example of the same variety:

$$\begin{array}{r} 47 \\ -23 \\ \hline \end{array}$$

Press 47

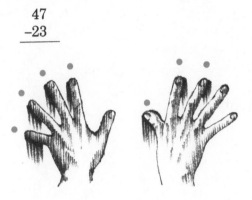

Clear (subtract) 23

Say "Ten"

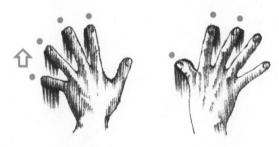

"Twenty"

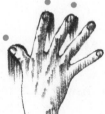

(Continue the count to "Twenty-Three" by Units)

"Twenty-One"

"Twenty-Two"

"Twenty-Three"

(5 to 4 exchange)

READ YOUR FINGERS.

The answer (difference) is 24.

Do just one more, being certain to count aloud as far as you are able by 10s and to continue your count to conclusion by Units. Ignore the amount which keeps changing on your fingers until you are ready to read your answer.

$$\begin{array}{r} 76 \\ -34 \\ \hline \end{array}$$

Press 76

Clear (subtract) 34

Say "Ten"

"Twenty"

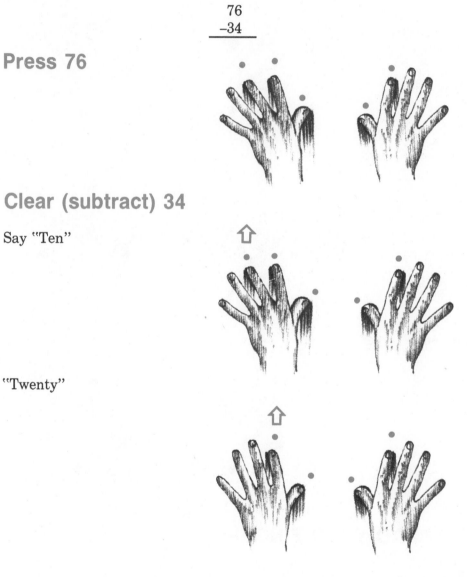

"Thirty"

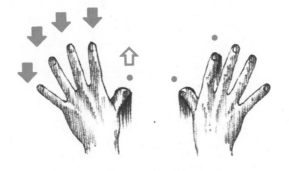

(50 to 40 exchange)

CONTINUE TO CLEAR BY UNITS

Say "Thirty-One"

"Thirty-Two"

(5 to 4 exchange)

"Thirty-Three"

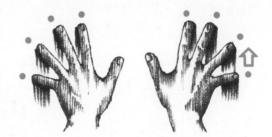

"Thirty-Four"

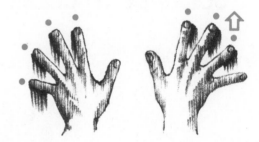

READ YOUR FINGERS.

The answer (difference) is 42.

Simple. And you can handle any subtraction of a single-digit or double-digit number from a double-digit minuend in exactly the same way.

Go on now to do the following page of examples, using this simple Fingermath method.

Subtraction Without Renaming

36 − 7	27 − 9	18 − 8	22 − 3	15 − 8
44 −10	37 −11	26 −15	32 −20	48 −33
50 −20	50 −13	50 − 6	50 −22	50 −31
60 −10	60 −14	63 −22	64 −35	65 ^s −16
73 −24	79 −38	76 −59	82 −47	81 −63
90 −46	96 −89	92 −18	91 −12	93 −47

The Unknown Quantity

Algebra: You Must Keep Your Balance

Does the word *algebra* turn you off, or even frighten you? There may be good reason. I suppose everyone is afraid of the dark, especially in unfamiliar surroundings. In algebra, the thought of trying to solve a problem without knowing all the facts is not unlike being in a dark room ... until one finds where the light switch is.

Part of that discovery is in the understanding of the word itself. Algebra is merely a term that refers to doing arithmetic with undefined letters or symbols in place of one or more of the definite numbers that usually appear in an example. Suppose you are calculating one of the examples in a previous section, such as:

$$3 + 2 + 5 = \square$$

Even in this simple, arithmetical form of addition, a symbol ($\square$) is used. You know that it stands for a number, and that the whole problem consists of finding what that number is. Then you enter that number, 10, inside the box. You have completed the mathematical statement that reads, "Three plus two plus five equals ten."

Now, switch the numbers around to new positions, like this:

$$3 + 2 + \square = 10$$

This sentence reads, "three plus two plus (How many?) equals ten." Instead of "How many?" you can put the letter *Y*, thus:

$$3 + 2 + Y = 10$$

That *Y* seems more ominous to lots of people than the empty box, but it's nothing but a short way of saying, "How many?" Suddenly it's algebra instead of arithmetic, but algebra is not necessarily higher mathematics. So whether you're using a □ or a *Y* or an *X* or any number-substitute, don't let it throw you. The way you change it back to a number is by keeping your balance.

Picture a balance beam and imagine that the fulcrum or point of support is the equals sign (=) in your Fingermath examples. All the numbers to the left of the equals sign belong on one end of the balance beam, and all the numbers to the right on the other end. What you've got to figure out is what will make both sides equal so that the beam will become level. In other words, you must end up with the same total on the left as you have on the right.

$$3 + 2 + \square = 10$$

Always work first with the side that has something missing, the side with a symbol or a letter.

PRESS 3

say "Three")

170

Then . . .

PRESS 2 MORE

say "One

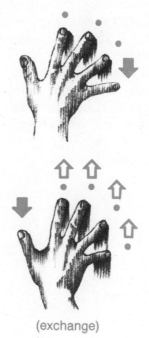

"Two")

(exchange)

(Keep your thumb pressed)

STOP

Now, you cannot add the □ , but you *can* add finger values until you end up with the number that appears on the other side of the equals sign, which is 10.

Begin your new count as usual with 1, and stop counting the moment you recognize the press of 10 on your fingers.

Here's your first real indication of how essential it is for you to master the skill of finger recognition; of knowing instantly, whether by sight or by feel, the value being pressed on either or both hands. In this example, you're counting from "one" to whatever number will have your fingers pressing 10. So your voice is saying one number while your fingers show another. The essential thing here is to recognize it in your fingers when you reach the final value you need, without getting confused by the numbers you're counting aloud.

That sort of skill is really quite common, if you stop and think about it.

You've often seen and heard performers playing piano or guitar and at the same time give all their attention, seemingly, to the words of the song they are singing. Perhaps you do it yourself, and think nothing of it. But if you do, you know it took a lot of practice.

Take this as a serious caution to spend some time at practice each day, responding to your helper's oral commands to press both single and double digit numbers. It's an exercise that pays off more and more as you get deep into computation.

Say "One

Say "Two

Say "Three

Say "Four

Say "Five"

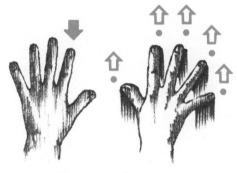

(9 to10 exchange)

THERE IT IS! 10!

In this case, you will have counted "Five" when you reach 10. So the number that makes your both sides balance is 5.

$$3 + 2 + \boxed{5} = 10$$

Once again, you can understand why I keep nagging you to practice recognition of finger values. If you can't automatically recognize it the moment your fingers read 10, you won't know when to stop counting.

Here's another one for practice:

18 + □ + 4 = 29

Just because the symbol □ appears between the 18 and the 4, that's no reason to get upset. Simply save the □ for last. Here goes:

PRESS 18

say "Eighteen"

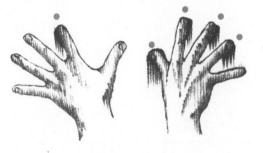

PRESS 4 MORE

say "One"

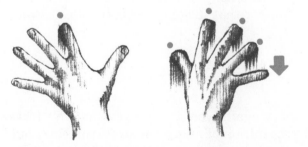

THE UNKNOWN QUANTITY

"Two"

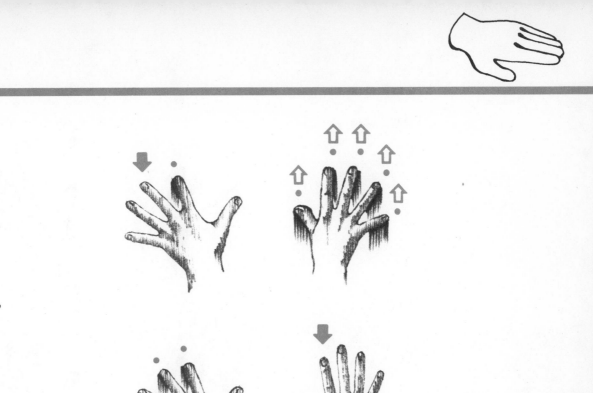

"Three"

"Four")

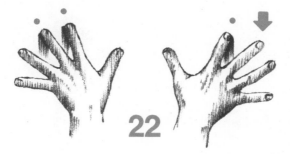

22

Your fingers now read 22. To make the balance beam level, they must read 29. Don't clear the press of 22, but begin your new count and stop when you reach 29.

Say "One"

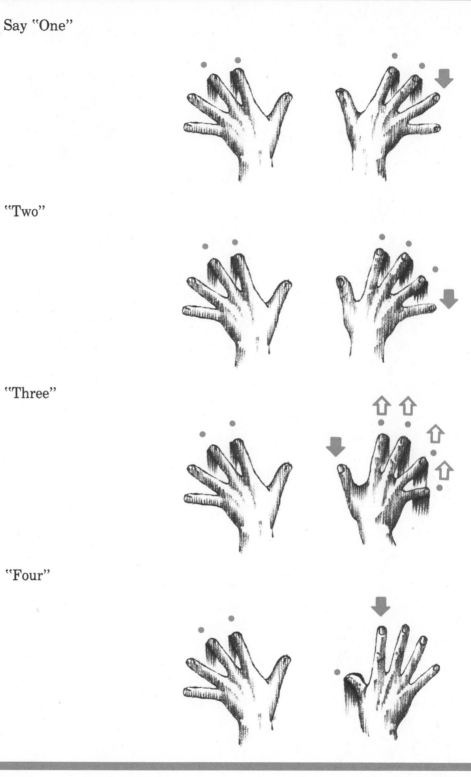

"Two"

"Three"

"Four"

"Five"

"Six"

"Seven"

29

THERE IT IS! 29!

The balance has equal values on each side. You've counted 7 to reach 29. 7 is the missing number.

Once you become accustomed to this easy procedure, start using the shortest possible route to your answer by employing the most direct manipulation.

In this last example (18 + ☐ + 4 = 29), after you reached your Press of 22, you could have avoided some of the unit-by-unit counting as follows:

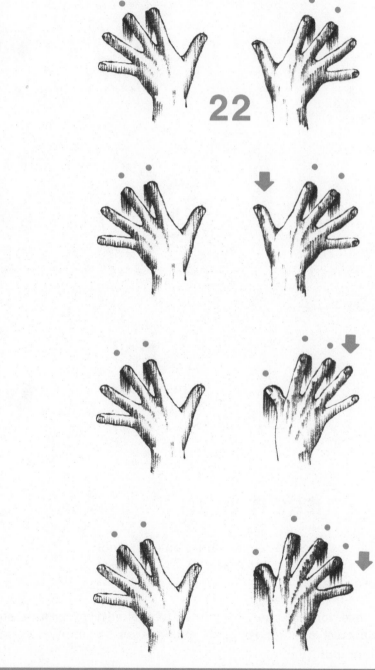

To reach 29:

say "Five"

"Six"

"Seven")

THE UNKNOWN QUANTITY

Much, much shorter. But remember that a child may have to use the longer route for some time to become familiar with the mechanics of reaching a number. You should revert to unit counting now and then yourself, for valuable practice that will serve you well in future manipulations.

Now go on to solve the algebra problems that follow. Remember to begin each new count with "One."

Finding The Unknown Number
Algebra Addition

$11 + \square + 11 =$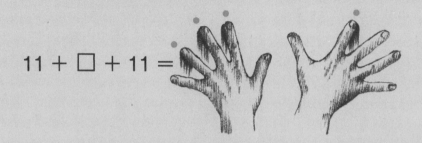

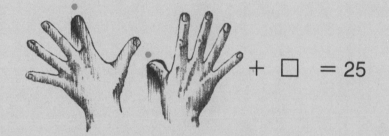 $+ \square = 25$

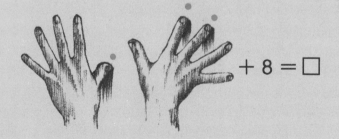 $+ 8 = \square$

$27 + 9 + \square = 52$ $99 - 8 - 9 - 8 = \square$

$13 + 2 + \square = 24$ $\square + 30 + 4 = 44$

$44 + Y + 9 = 64$
$Y = \square$ $\square + 6 + 9 = 36$

$4 + 1 + \square + 12 + 3 = 30$

$71 + N + 8 + 9 = 99$ $Y = 7$
$N = \square$ $12 + \square + Y = 29$

$16 + 4 + 5 + \square = 46$ $12 - 3 - 3 = \square$

More Missing Addends (Algebra)

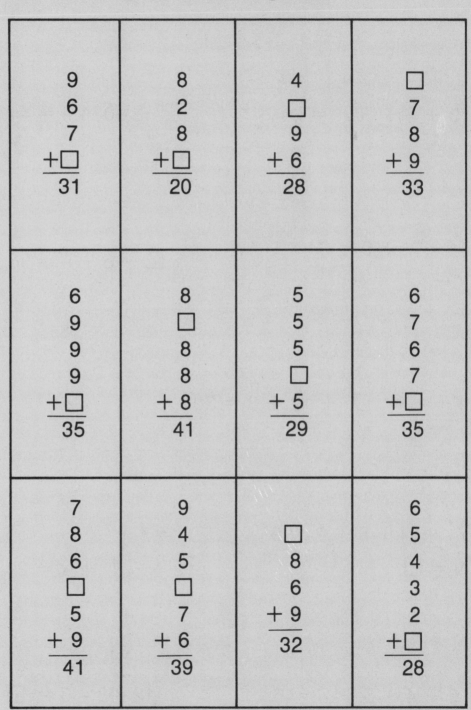

9 6 7 + □ ___ 31	8 2 8 + □ ___ 20	4 □ 9 + 6 ___ 28	□ 7 8 + 9 ___ 33
6 9 9 9 + □ ___ 35	8 □ 8 8 + 8 ___ 41	5 5 5 □ + 5 ___ 29	6 7 6 7 + □ ___ 35
7 8 6 □ 5 + 9 ___ 41	9 4 5 □ 7 + 6 ___ 39	□ 8 6 + 9 ___ 32	6 5 4 3 2 + □ ___ 28

SUBTRACTION EQUATIONS

Put this same technique to work to discover the missing number in equations that involve subtraction.

$$19 - 3 - \square = 11$$

Start, as before, on the side that has the missing value.

PRESS 19

(Say "Nineteen")

Then, subtract 3

(Say "One-Two-Three")

Now, you cannot subtract $\square$ but you can subtract finger values until you end up with the number (11) that appears on the other side of the equals sign.

Begin your new count, as always, with "One" and stop counting the moment you recognize the press of 11.

You can use a unit-by-unit count (being certain to begin with "One"). However, take a good look and you'll see that you can simply clear your right thumb (saying "Five") to achieve a press of 11.

Say "One"

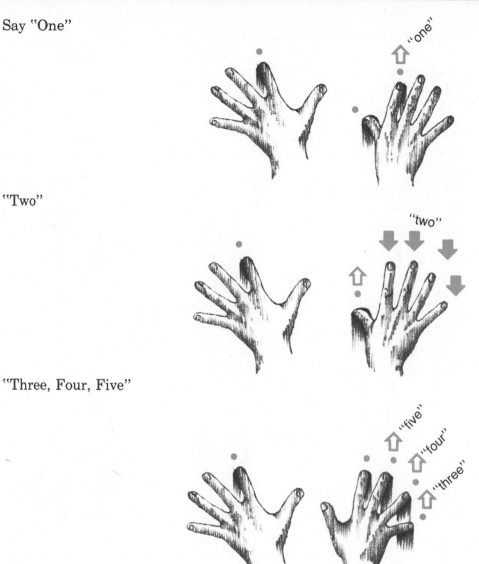

"Two"

"Three, Four, Five"

There it is!

The balance is level.

You've cleared 5 to reach 11. 5 is the missing number, the required answer to the problem.

Go on now to solve the problems on the following pages. There is no longer any excuse for letting yourself be scared off by letters or symbols. Meet them head on!

Algebra Subtraction

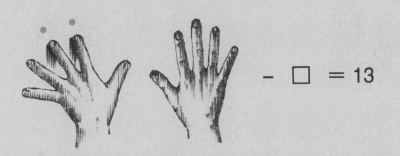 $-\ \square\ = 13$

$18 - 6 - \square = 4$ $27 + 9 - \square = 27$

$6 - \square = 2$ $36 - 8 - 4 - \square = 15$

$27 + 8 - \square = 27$ $8 + 8 - \square = 11$

$-\ \square\ = 26$

$39 - X = 23$

$X = \square$

$15 - \square + 30 = 41$

$60 + Y - \square = 57$

$Y = 8$

$93 - 8 - 8 - 8 = \square$

$48 + 6 - Z = 44$

$Z = \square$

$22 + 4 + 9 - \square = 20$

$36 - 2 + 8 - 5 = \square$

$24 + 9 - 8 = \square$

$36 - 9 - 9 - 9 - 9 = \square$ $83 - 6 + 3 - \square = 71$

Multiplication

Repeated Addition

Examine any simple multiplication example, paying close attention to the way it's written. You'll quickly realize something you may never have noticed before, something that takes all the anxiety out of the prospect of solving the problem. The fact is, multiplication is nothing more than addition. Repeated addition.

Read this aloud:

$$8 \times 2 = \square$$

"Eight times two equals how many?"

You could change this sentence without disturbing either side of the equation, and say, "The number 2 added eight times equals how many?" Literally, this is the meaning of those numbers and symbols, and when you look at them that way, they are easier to comprehend and a lot easier to work with. Repeated addition is especially convenient and simple with Fingermath, because—as you're seeing more and more clearly—your fingers have the ability to accumulate and store an endless series of numbers. When every one of these numbers happens to be the same, you *could* say you're multiplying. Best of all, when you employ addition in order to multiply, it becomes unnecessary to memorize multiplication tables . . . and many of us have suffered enough through *that* project.

Now don't misquote me as saying that you shouldn't know those tables. The point is that this knowledge can better be gained through regular Fingermath practice than by rote. It's desirable to master the tables, and know them by heart. But if you or your child has been stymied by this or that table, forget it for now because you have your ten-finger calculator at hand. You can multiply by adding with Fingermath.

BASIC LEVEL FIRST

The best and clearest route to every advanced Fingermath technique is via unit-by-unit counting. Multiplication is no exception; in fact, the logic of this procedure is made especially clear, for both adults and children, by means of unit-by-unit counting.

As you begin to recognize a pattern in the repeated unit-counting of any given number, you will also be getting the *feel* of that number and begin to sense the unique way it fits under your fingers. You will discover repeated sequences that fall into characteristic patterns for each of the numbers.

I am starting with multiplication of the number 1. You may think that's the obvious thing to do, but I won't be following the obvious, natural order, going on to 2 and 3 and on up. I'm going to jump around and introduce you to each of the right-hand numbers in the order that is easiest for you to follow; that is, in the order of increasing difficulty:

1 – 2 – 5 – 6 – 9 – 8 – 4 – 7 – 3

Your first step will be to concentrate on unit-counting each number individually and repeatedly, ten times (10 X ☐).

 10 X 1

You've already counted to 99 by units, so I'm not going to waste your time here with diagrams. Up to now, the instructions were: "Press 1 . . . and 1 more . . . and 1 more. . . . How many?" Then you read your fingers for a total. In multiplication, you've got to count the *number of times* you are pressing a particular number—in this instance, 1.

Press 1s now, counting aloud at each press until you say "ten."

"One – Two – Three – Four – Five – Six – Seven – Eight – Nine – Ten"

Now read your fingers.

You have carried out the operation 10×1
by adding the number 1 ten times.

$$10 \times 1 = 10$$

2s

Learning how to finger your way through the multiplication of numbers other than 1, you will start with the most familiar technique and conclude with the most advanced and direct.

In basic Fingermath, it's obvious that every 2 is made up of the count "One – Two," whether you are counting aloud or not. Again, you will rely on this basic unit-by-unit structure to discover the route to the repeated addition (multiplication) of 2s. Before paying any attention to how many 2s you are pressing, however, you must first observe where your fingers end up each time you count off another 2. Try it now, following each drawing and talking your way unit by unit through this sequence of 2s.

PRESS 2 (finger-by-finger)

Say "One – Two"

Read your fingers (2)

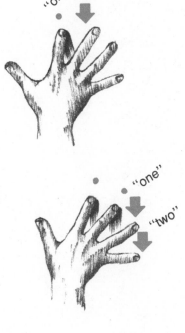

PRESS 2 MORE

Say "One – Two"

Read your fingers (4)

Remember, you are not paying attention just yet to how many 2s you are accumulating, but you are noting the look and feel of your fingers each time you complete a new 2.

PRESS 2 MORE

Say "*One . . .*

"Two"

Read your fingers (6)

PRESS 2 MORE

Say "One - Two"

Read your fingers (8)

PRESS 2 MORE

Say "One" . . .

"*Two*"

Read your fingers (10)

PRESS 2 MORE

Say "One - Two"

STOP! (KEEP YOUR FINGERS PRESSED.)

Notice what's happening? Your right hand is now starting to repeat the entire sequence that has just come to an end. And it will keep on repeating it. Watch carefully and you'll see this repetition.

Now read your fingers (12)

PRESS 2 MORE

Say "One – Two"

Read your fingers (14)

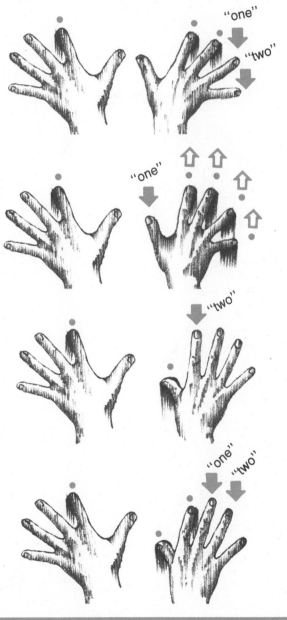

PRESS 2 MORE

Say *"One* . . .

"Two"

Read your fingers (16)

PRESS 2 MORE

Say "One – Two"

Read your fingers (18)

(Notice how your right hand has repeated the 2–4–6–8 fingering)

PRESS 2 MORE

Say "One" . . .

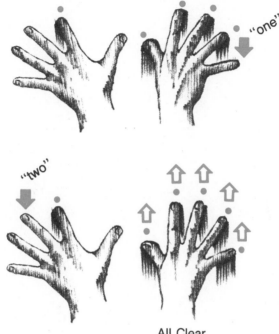

"Two"

Read your fingers (20)

All Clear

STOP AND HOLD ONTO IT!

STOP!

Now you're ready to repeat the sequence all over again. On your right hand you will always end up on a 2, a 4, a 6, or an 8, and on your left hand you will start accumulating 10s. Try going on, using this basic unit-by-unit counting of 2s. See if you can go up to 98.

Then start all over again, this time not stopping to read your fingers, but forcing yourself into a steady pressing of 2s with a rhythm that is set by your voice counting "One–Two . . . One–Two . . . One–Two," all the way to 98.

If you're working with a child, go only as far as the youngster can count and can recognize finger values. Even going only to 10 is fine for a start. Then advance to 20. Take your time. Repetition is the key.

When you feel really confident pressing 2s at this basic level, you're ready to start multiplying 2s. Use the identical finger manipulations but change your voice accompaniment. You are now going to count each completed 2.

Follow the next set of drawings and say the count as indicated.

(2)

Press 2s, unit-by-unit, counting how many
2s you complete.

PRESS 2   (by units, remember)

Say "One 2"

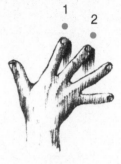

PRESS 2

Say "Two 2s"

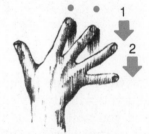

PRESS 2

Say "Three 2s"

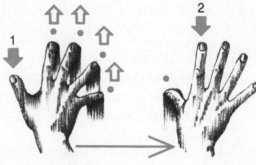

(exchange)

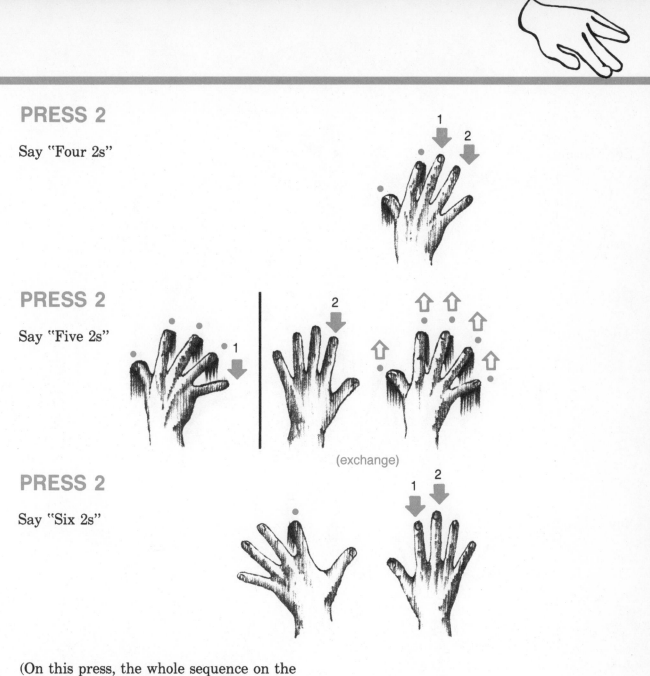

PRESS 2

Say "Four 2s"

PRESS 2

Say "Five 2s"

(exchange)

PRESS 2

Say "Six 2s"

(On this press, the whole sequence on the right hand begins again. Observe your fingers—*feel* them—each time you complete the press of another 2.)

PRESS 2

Say "Seven 2s"

PRESS 2

Say "Eight 2s"

(exchange)

PRESS 2

Say "Nine 2s"

PRESS 2

Say "Ten 2s"

(exchange)

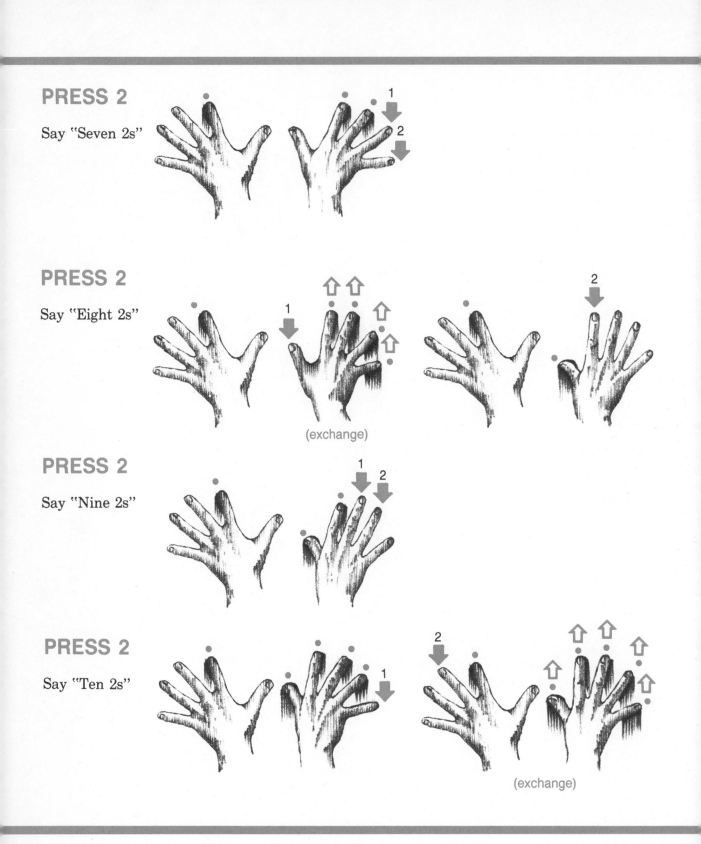

196

Now read your fingers:	20
2 added 10 times =	20
10 X 2 =	20

Repeat this entire sequence four or five times. At first, pause to observe your fingers after each new set of 2 is established. Note again that, following the fifth 2, your fingers repeat the identical pattern of manipulations. If you are able, go to 98.

Then repeat the entire sequence three times, not stopping to read your fingers but maintaining an even pace, keeping in time with your voice as you call out how many 2s you are pressing.

But enough of this unit-by-unit counting of 2s! You're ready now to move onto the fast track. Here's your new sequence of manipulations, which promise the speediest and most direct route from one press of 2 to the next. You'll see at once how your unit counting has brought you smoothly to the stage of adding each 2 with a single press of the whole number.

For the moment, don't try to count how many times you are pressing. Concentrate only on pressing each 2 in a single operation, and observing how your fingers end up each time. *Never* count out the accumulating total stored on your fingers. For starters, pay attention only to the pattern of 2s, pressing each as a whole number. Counting will come afterwards.

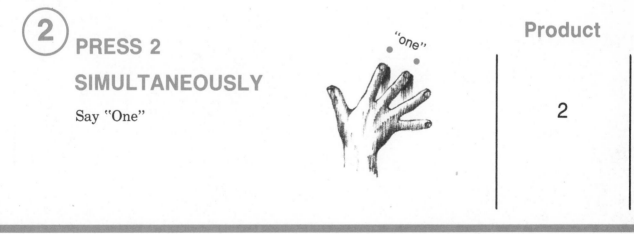

2 PRESS 2 SIMULTANEOUSLY

Say "One"

"one"

Product

2

		Product

X2   **PRESS 2 MORE SIMULTANEOUSLY**

Say "Two"

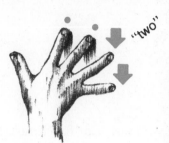

4

X3   **PRESS 2 MORE SIMULTANEOUSLY**

Say "Three"

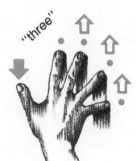

6

(Your index finger remains pressed as you simultaneously press your thumb and clear your other fingers. As an adult, you happen to know 5 − 3 = 2, and that's the manipulation here, isn't it? You're simply pressing 5 and clearing 3.)

X4   **PRESS 2 MORE SIMULTANEOUSLY**

Say "Four"

8

		Product
X5	**PRESS 2 MORE SIMULTANEOUSLY**	
	Say "Five"	10

"five"

(Note that your pinky is not in action; it's bypassed in manipulating the instant press of 2, from 8 to 10. Now your entire right hand is free to repeat again the same sequence of 2s.)

X6	**PRESS 2 MORE SIMULTANEOUSLY**	
	Say "Six"	12

"six"

X7	**PRESS 2 MORE SIMULTANEOUSLY**	
	Say "Seven"	14

"seven"

2 S

X8 **PRESS 2 MORE SIMULTANEOUSLY**

Say "Eight"

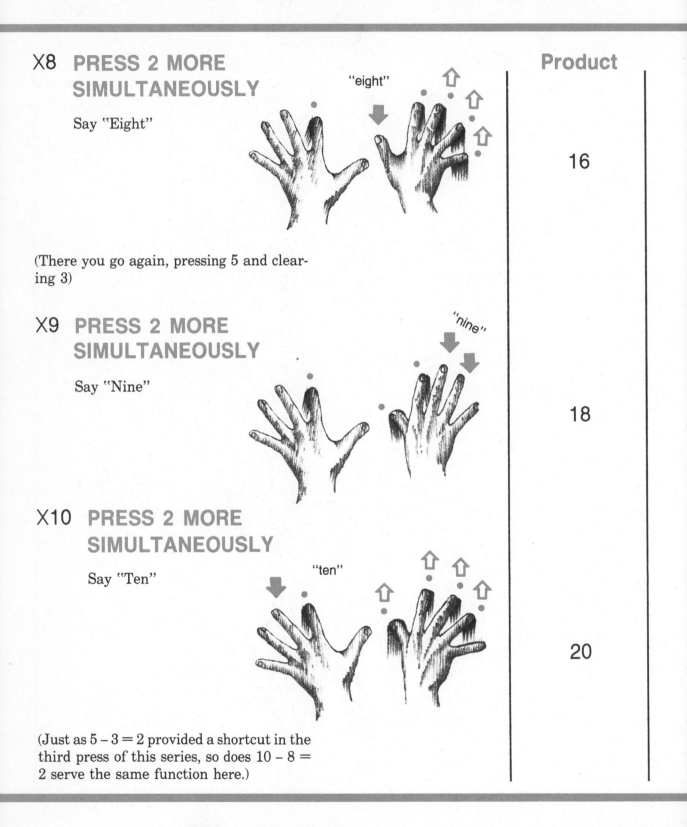

"eight"

Product

16

(There you go again, pressing 5 and clearing 3)

X9 **PRESS 2 MORE SIMULTANEOUSLY**

Say "Nine"

"nine"

18

X10 **PRESS 2 MORE SIMULTANEOUSLY**

Say "Ten"

"ten"

20

(Just as $5 - 3 = 2$ provided a shortcut in the third press of this series, so does $10 - 8 = 2$ serve the same function here.)

Repeat the entire series again and then again, until you can press sets of 2s in sequence without faltering.

You picked up some valuable information on the way. You saw that, when you didn't have two adjacent fingers available for a press, there was a rapid way of jumping over the necessary exchange.

If the right thumb was clear, you could press 5 and clear 3—a manipulation that works because $5 - 3 = 2$. Similarly, if the thumb was not clear, you could add 2 instantaneously by pressing 10 and clearing 8.

Three Ways to Add 2

Press 2

Press 5 and clear 3

Press 10 and clear 8

Like all exchanges, these must be performed as a simultaneous manipulation. Don't ever press the 5 first and then clear 3. Work them together.

After you've pressed through the entire sequence at least five times, see if you can press 2s in this same sequence, but now count the number of times you press. Go as far as you can go, ending up with a product of 98. (I still haven't shown you how to break 100, but I'm saving that plum for later.)

Here is a series of exercises that will reinforce your ability to press 2s repeatedly when they do not occur in the now familiar sequence which results from an initial press of 2. You still will use one of the established ways of pressing each 2.

First, Press 1. Then press 2s ten times.
When finished, your fingers read 21.
Next, Press 3. Then press 2s ten times.
When finished, your fingers read 23.
Next, Press 4. Then press 2s ten times.
When finished, your fingers read 24.

Continue this pattern, first pressing each of the numbers up to 9, followed by ten consecutive 2s. Note that at the end of each series of ten 2s you easily can recognize the correct result because the number first pressed clearly is part of the final sum.

When you feel truly secure pressing 2s, and you can maintain a steady pace, without hesitating, as you rhythmically count aloud the number of times you are pressing, go on to the following page of examples. Do all of them (not necessarily at one sitting) both in writing and orally, having someone read each one to you.

Then, and only then, you'll be ready for the next multiplication series.

Multiplying 2s

This page to be done orally also.

1 X 2 = 2 X 2 = 3 X 2 =	4 X 2 = 5 X 2 = 6 X 2 =	7 X 2 = 8 X 2 = 9 X 2 =	10 X 2 = 0 X 2 = 2 X 2 =	4 X 2 = 6 X 2 = 8 X 2 =
10 X 2 = 1 X 2 = 3 X 2 =	5 X 2 = 7 X 2 = 9 X 2 =	10 X 2 = 8 X 2 = 6 X 2 =	4 X 2 = 2 X 2 = 9 X 2 =	7 X 2 = 5 X 2 = 3 X 2 =
11 X 2 = 12 X 2 = 13 X 2 =	14 X 2 = 15 X 2 = 16 X 2 =	17 X 2 = 18 X 2 = 19 X 2 =	20 X 2 = 16 X 2 = 12 X 2 =	18 X 2 = 14 X 2 = 12 X 2 =
19 X 2 = 17 X 2 = 15 X 2 =	13 X 2 = 11 X 2 = 14 X 2 =	8 X 2 = 10 X 2 = 6 X 2 =	14 X 2 = 4 X 2 = 16 X 2 =	6 X 2 = 18 X 2 = 2 X 2 =
3 X 2 = 12 X 2 = 5 X 2 =	13 X 2 = 4 X 2 = 15 X 2 =	11 X 2 = 7 X 2 = 5 X 2 =	16 X 2 = 17 X 2 = 19 X 2 =	4 X 2 = 6 X 2 = 18 X 2 =
0 X 2 = 9 X 2 = 1 X 2 =	11 X 2 = 12 X 2 = 14 X 2 =	3 X 2 = 5 X 2 = 2 X 2 =	8 X 2 = 7 X 2 = 6 X 2 =	10 X 2 = 4 X 2 = 9 X 2 =

5s

Your experience with 2s has put you on the road to understanding and mastering all the other multiplication manipulations. Now you're going to tackle 5s. Once again, by first observing where your fingers land each time you complete the unit count of a new 5, you will quickly see a pattern that keys you into the advanced manipulation of pressing whole numbers. Because you already are experienced at pressing the right thumb as 5, whenever it is available, you may continue to do this. Elsewhere, press 5s by units.

(5) **PRESS 5 DIRECT**

Say "Five"

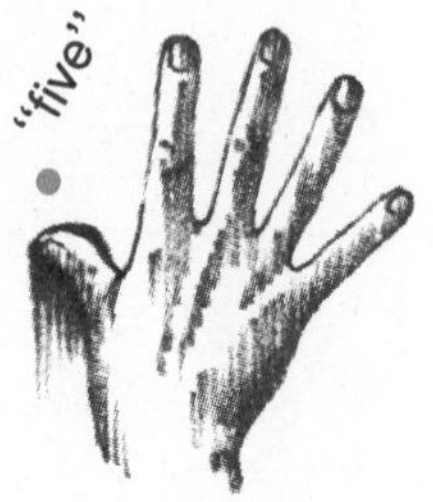

Now PRESS 5 Units

Say "One, Two, Three, Four,

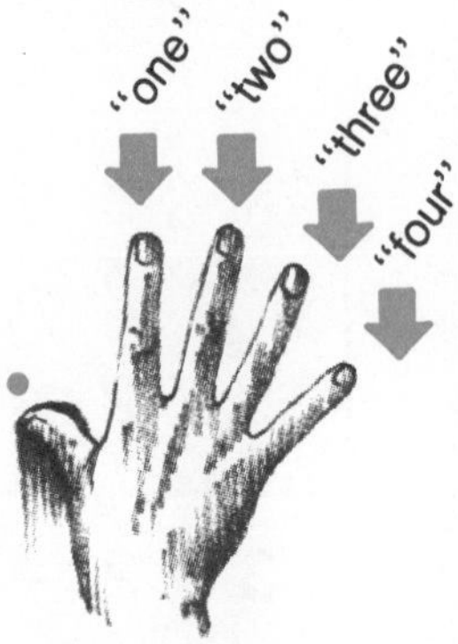

Five"

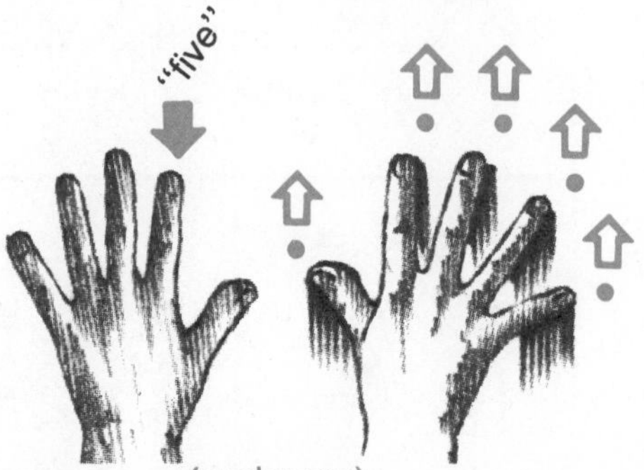

(exchange)

PRESS 5 DIRECT

Say "Five"

PRESS 5 units

Say "One, Two, Three, Four,

Five"

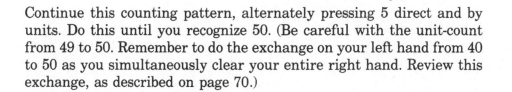

(exchange)

Continue this counting pattern, alternately pressing 5 direct and by units. Do this until you recognize 50. (Be careful with the unit-count from 49 to 50. Remember to do the exchange on your left hand from 40 to 50 as you simultaneously clear your entire right hand. Review this exchange, as described on page 70.)

Repeat the whole 5s sequence at least three times more, observing how the addition of each 5 ends with your fingers pressed in one of three ways:

Right thumb exclusively (first press only)

10s finger(s) exclusively

10s finger(s) plus Right Thumb

You never end with any fingers pressed on your right hand except the thumb.

Pay no attention yet to the number of times you're pressing a 5. Concentrate *only* on the way each 5 concludes.

Using your new awareness of how these 5s fall into place, go through the sequence again, this time dispensing with units altogether. Continue to ignore your finger totals. Concentrate on this short and simple method of making exchanges:

When your right thumb is available, press 5 direct.

When it is not available, press the next available 10s finger and simultaneously clear your right thumb. (While the child may not, *you* know that 10 − 5 = 5. That's the 5 Exchange.)

Here it is; give it a try. Then repeat it ten times before going on:

PRESS 5 DIRECT

Say "Five"

PRESS 5 MORE (EXCHANGE)

Say "Five"

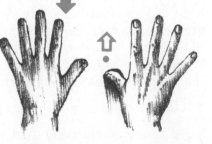

206

PRESS 5 MORE DIRECT

Say "Five"

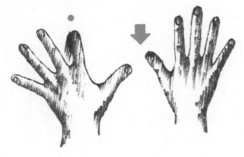

PRESS 5 MORE
(EXCHANGE)

Say "Five"

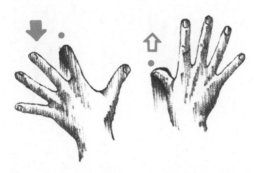

Keep going until you recognize 50. Be sure to go through this entire sequence ten times. Try going all the way to 95, steadily calling out Five-Five-Five. . . . Keep an even pace and practice until you can press successive 5s automatically.

See the pattern? Can you feel it? It's so simple, even the littlest child picks it up quickly. Continue to direct press and exchange 5s alternately until you recognize 50. Pay no attention to the total that is accumulating. Just get to feel comfortable with this series of manipulations while you move along at an even pace—not fast, just steady.

Now at last you're ready to press 5s repeatedly as the route to multiplication. For this following sequence to 50, I've given you the entire set of diagrams. Now you must aim at getting used to counting the number of times you are pressing 5.

5 S

207

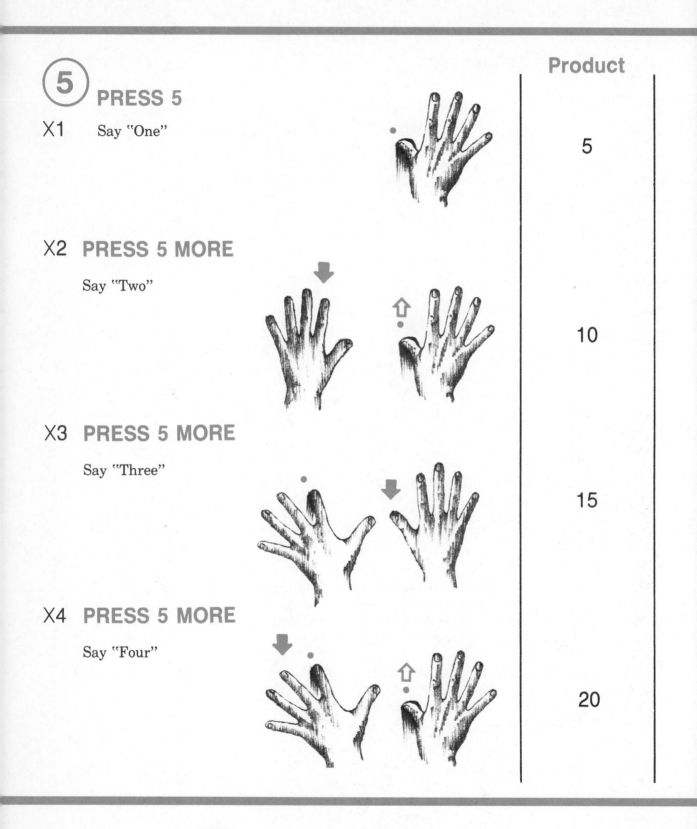

		Product
(5) **PRESS 5**		
X1 Say "One"		5
X2 **PRESS 5 MORE** Say "Two"		10
X3 **PRESS 5 MORE** Say "Three"		15
X4 **PRESS 5 MORE** Say "Four"		20

208

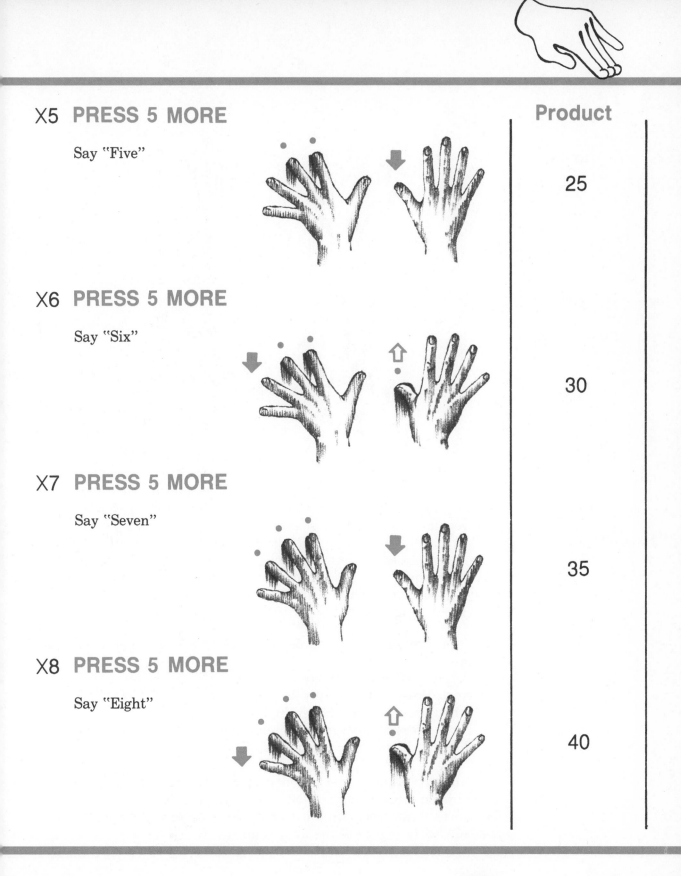

		Product
X5 **PRESS 5 MORE** Say "Five"		25
X6 **PRESS 5 MORE** Say "Six"		30
X7 **PRESS 5 MORE** Say "Seven"		35
X8 **PRESS 5 MORE** Say "Eight"		40

X9 PRESS 5 MORE

Say "Nine"

X10 PRESS 5 MORE

Say "Ten"

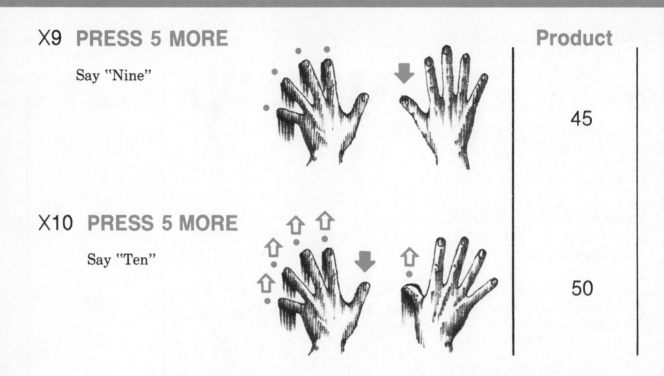

Product

45

50

If you really apply yourself you can move right on to 95. Try it with your eyes closed. And do it to some music that will establish a steady rhythm for you to follow. Don't be satisfied until you (or your child) can perform with regularity—not fast, just evenly.

Here is a series of exercises that will reinforce your ability to press 5s repeatedly when they do not occur in the now familiar sequence which results from an initial press of 5. You still will use one of the established ways of pressing each 5.

First, Press 1. Then press 5s ten times. (Your right index remains pressed.)
When finished, your fingers read 51.
Next, Press 2. Then press 5s ten times.
When finished, your fingers read 52.
Next, Press 3. Then press 5s ten times.
When finished, your fingers read 53.

Continue this pattern, first pressing each of the numbers up to 9, followed by ten consecutive 5s. Note that at the end of each series of ten 5s you easily can recognize the correct result because the number first pressed clearly is part of the final sum.

210

Multiplying 2s and 5s

This entire page to be done orally also.

$2 \times 2 =$ $2 \times 5 =$ $3 \times 2 =$	$3 \times 5 =$ $4 \times 2 =$ $4 \times 5 =$	$5 \times 2 =$ $5 \times 5 =$ $6 \times 2 =$	$6 \times 5 =$ $7 \times 2 =$ $7 \times 5 =$	$8 \times 2 =$ $8 \times 5 =$ $9 \times 2 =$
$9 \times 5 =$ $10 \times 2 =$ $10 \times 5 =$	$11 \times 2 =$ $11 \times 5 =$ $12 \times 2 =$	$12 \times 5 =$ $13 \times 2 =$ $13 \times 5 =$	$14 \times 2 =$ $14 \times 5 =$ $15 \times 2 =$	$15 \times 5 =$ $16 \times 2 =$ $16 \times 5 =$
$1 \times 2 =$ $1 \times 5 =$ $10 \times 2 =$	$0 \times 5 =$ $9 \times 2 =$ $1 \times 5 =$	$8 \times 2 =$ $2 \times 5 =$ $7 \times 2 =$	$3 \times 5 =$ $6 \times 2 =$ $4 \times 5 =$	$5 \times 2 =$ $5 \times 5 =$ $4 \times 2 =$
$6 \times 5 =$ $3 \times 2 =$ $7 \times 5 =$	$2 \times 2 =$ $8 \times 5 =$ $1 \times 2 =$	$9 \times 5 =$ $0 \times 2 =$ $10 \times 5 =$	$20 \times 2 =$ $11 \times 5 =$ $19 \times 2 =$	$12 \times 5 =$ $18 \times 2 =$ $13 \times 5 =$
$17 \times 2 =$ $14 \times 5 =$ $16 \times 2 =$	$15 \times 5 =$ $15 \times 2 =$ $16 \times 5 =$	$14 \times 2 =$ $17 \times 5 =$ $13 \times 2 =$	$18 \times 5 =$ $12 \times 2 =$ $19 \times 5 =$	$11 \times 2 =$ $0 \times 5 =$ $10 \times 2 =$
$1 \times 5 =$ $8 \times 2 =$ $3 \times 5 =$	$6 \times 2 =$ $5 \times 5 =$ $4 \times 2 =$	$7 \times 5 =$ $2 \times 2 =$ $9 \times 5 =$	$0 \times 2 =$ $11 \times 5 =$ $1 \times 2 =$	$10 \times 5 =$ $3 \times 2 =$ $9 \times 5 =$
$4 \times 2 =$ $8 \times 5 =$ $5 \times 2 =$	$7 \times 5 =$ $6 \times 2 =$ $6 \times 5 =$	$7 \times 2 =$ $5 \times 5 =$ $8 \times 2 =$	$4 \times 5 =$ $9 \times 2 =$ $3 \times 5 =$	$10 \times 2 =$ $2 \times 5 =$ $11 \times 2 =$

A DETOUR

There are some important applications of this 5 exchange that come into play in manipulations for addition. Even though you're in the midst of multiplication, this is the best time to take a close look at them, while the 5 exchange is fresh in your mind.

I've set up a couple of workpages of addition examples that will quicken your perception of this critical exchange and the situations in which you should—and should not—employ it. To help you through these exercises, I have used two different symbols wherever a 5 appears. A typical example looks like this:

Every 5 with this symbol ■ requires a direct press of your right thumb. Every 5 with this symbol ▲ requires the 5 exchange. Whenever you see a 2, try to press the whole number. For numbers other than 5 and 2, use only unit counting at present. Later, when you've practiced enough, your fingers will know what to do instinctively!

To use this technique successfully you must be able to hold onto established fingers. Look at the example above and press through it as follows:

⑤ PRESS 5 DIRECT

Say "Five"

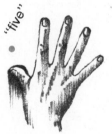

+ **(3)** **PRESS 3 MORE**

Say "One–Two–Three"

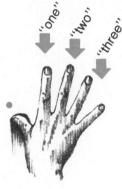

+ **(5)** **PRESS 5 MORE (EXCHANGE)**

Say "Five"

You are exchanging only the 5. Keep the three fingers pressed.

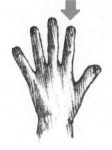

+ **(5)** **PRESS 5 MORE DIRECT**

Say "Five"

Hold onto that 3!

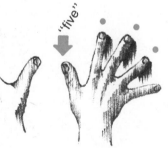

+ **(2)** **PRESS 2 MORE DIRECT**

Say "Two"

Read your fingers (20)

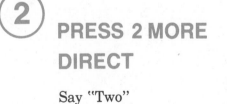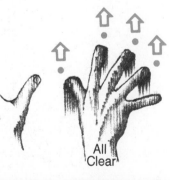

All Clear

You'll get plenty of practice holding onto established fingers as you do the examples on the following pages. This same technique will also come into play with subtraction, multiplication, and division. You must simply remember to press or to clear *only* the required value.

Go very slowly. Think through each manipulation carefully. Let your brain lead your fingers. Eventually, your fingers will seem to be doing the thinking themselves.

Some of these number sequences are *designed* to be difficult, to make you exercise every possible combination of fingers. Whenever you complete a tough one, go back immediately and repeat it a few times.

Pressing 5s - Direct and with Exchanges - Addition

Press 2s directly as whole numbers

This entire page to be repeated orally.

5■ 4 4 + 5■	5■ 2 2 + 2	2 5■ 4 + 5■	5■ 3 2 + 5■	5■ 4 1 + 5■	1 5■ 2 + 5▲	5■ 2 4 + 5■
5■ 5▲ 2 + 5■ ⌐	4 1 2 + 5▲	3 5■ 5▲ + 4	5■ 2 5▲ + 2	4 1 5▲ + 5■	1 5■ 5▲ + 2	2 2 5■ + 5▲
2 3 5▲ + 5■	2 2 2 + 5▲	2 1 4 + 5▲	2 5■ 2 + 5▲	5■ 2 5▲ + 2	5■ 3 2 + 5■	6 5▲ 2 + 5■
1 1 4 + 5▲	4 3 2 + 5▲	3 4 5▲ + 5■	4 2 5▲ + 2	3 3 3 + 5▲	5■ 5▲ 5■ + 5▲	2 5■ 5▲ + 2
5■ 2 5▲ 5■ + 2	6 5▲ 2 2 + 5▲	5■ 3 5▲ 2 + 5▲	5■ 5▲ 5■ 5▲ + 5■	2 5■ 2 5▲ + 2	5■ 5▲ 2 5■ + 5▲	7 5▲ 5■ 2 + 5▲

Pressing 5s - Direct and with Exchanges

Press 2s Directly As Whole Numbers

6 5 2 5 5 + 2	5 5 5 2 2 + 5	2 2 5 2 5 + 5	1 2 5 3 5 + 2	6 2 5 5 2 + 5	9 5 2 5 6 + 5	2 2 2 5 5 + 5
8 5 5 2 4 + 2	7 2 5 5 5 + 5	5 6 5 5 2 + 5	4 2 5 2 5 + 2	9 5 9 5 2 + 5	3 2 5 5 2 + 2	8 5 2 2 5 + 5
5 6 5 5 2 5 + 5	6 2 5 2 5 5 + 5	8 2 5 5 2 2 + 5	9 1 5 2 1 2 + 5	5 4 1 5 5 2 + 5	7 2 1 2 5 5 + 5	5 5 5 2 5 5 + 5

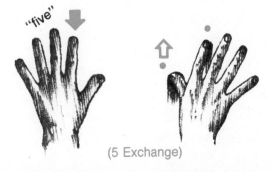

6s

Many pages back, when you first met with the direct press for 5, you also learned a quick way to achieve a 6. You merely had to press 5 and then add a unit count to make 6. At the time, however, you could do this only when the right thumb was available; otherwise, you used unit counting entirely.

Now that you've got 5s under your belt—both direct and via the exchange—you will use these same maneuvers to execute the repetition of 6s. Either press 5 direct and add 1 more, or do a 5 exchange and add 1 more.

Now, in order to get into the swing of this series, you've first got to take some double steps. Pay close attention to these instructions and diagrams. They're not at all difficult because they're so closely related to 5s. At this point, ignore the totals you're accumulating; concentrate *only* on how to manipulate each 6.

 PRESS 6

   say "Six"

(Whenever the right hand is all clear, you can press 6 as a whole number.)

"six"

PRESS 6 MORE

say "Five"

(Don't clear the right index finger)

"five"

(5 Exchange)

"Six"

(Don't clear the right index finger)

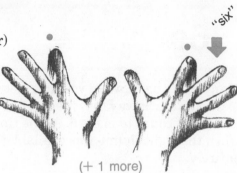

(+ 1 more)

PRESS 6 MORE

Say "Five – Six"

(Don't clear the right index
or middle finger)

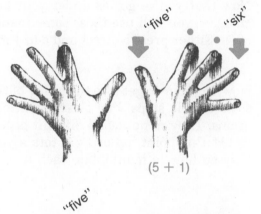

(5 + 1)

PRESS 6 MORE

Say "Five"

(Don't clear the right index, middle,
or ring finger)

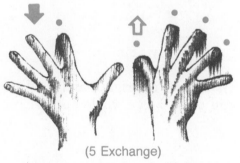

(5 Exchange)

"Six"

(Don't clear the right index,
middle, or ring finger)

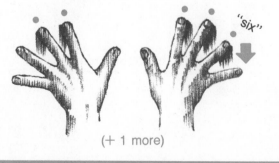

(+ 1 more)

218

PRESS 6 MORE

Say "Five"

(Hold onto all four fingers)

"Six"

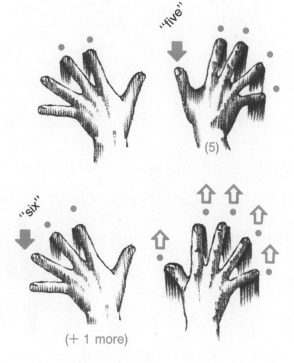

"five"

(5)

"six"

(+ 1 more)

At this point, your right hand has been entirely cleared and the whole sequence begins all over again. You can press the next 6 as a whole number and then follow the established pattern of 5 plus 1 more. Continue until you recognize 60.

Do the entire sequence again, with the diagrams if necessary. Then try it on your own, concentrating on the simple construction of 6s; think through every 5 plus 1 more. Next, do it with your eyes closed, and to music.

Now you're going to take your newly acquired skill and use it to press 6s repeatedly as whole numbers. You still must gear your mind to think the 5 plus 1 combinations; only this time around, do not press until your head signals your fingers to execute the appropriate instant press.

Again ignore your totals; just concentrate on executing clean, steady 6s.

PRESS 6 DIRECT

Say "Six"

PRESS 6 MORE
(EXCHANGE + 1)

Say "Six"

PRESS 6 MORE DIRECT

Say "Six"

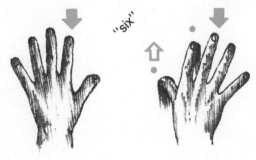

PRESS 6 MORE
(EXCHANGE + 1)

Say "Six"

220

PRESS 6 MORE

Although you've been practicing this as a 5 plus 1 maneuver, there's another way of looking at it. Consider 10 minus 4. That's the way to make a direct press with a one-step thought. Watch.

You've got 24 pressed

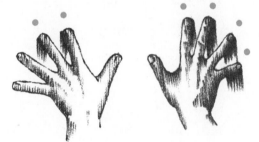

PRESS 6 MORE

(Press 10, clear 4)

Say "Six"

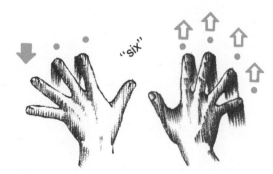

Now you can repeat the same sequence starting with a direct press of 6.

Continue this pressing of 6s to 60. Then go through it again as far as you can until all your fingers are used up!

6 S

You must go back and press this entire sequence over and over, until you are able to execute each successive manipulation cleanly and unhesitatingly. If you proceed without having acquired this skill, you'll do yourself a disservice. Remember that speed is of little consequence, so don't rush into a new challenge until you feel comfortable with the one you're working on. There's so little to learn in Fingermath, as you'll see, that even when you take the time to practice every new manipulation religiously, you'll complete the entire picture before you know it!

In order to apply your new skill in repeating 6s to multiply, you will execute the same sequence, counting the number of times you press 6. Remember that you have now learned three possible ways of achieving a 6. Here's the picture:

Three Ways to Press 6

5 + 1 (Right Hand)

5 (Exchange) + 1

10 − 4

Here's the complete series of 6s up to 60. Learn it!

Product

 PRESS 6

✕ 1   Say "One"

6

	Product
×2 PRESS 6 MORE Say "Two"	12
×3 PRESS 6 MORE Say "Three"	18
×4 PRESS 6 MORE Say "Four"	24
×5 PRESS 6 MORE Say "Five"	30

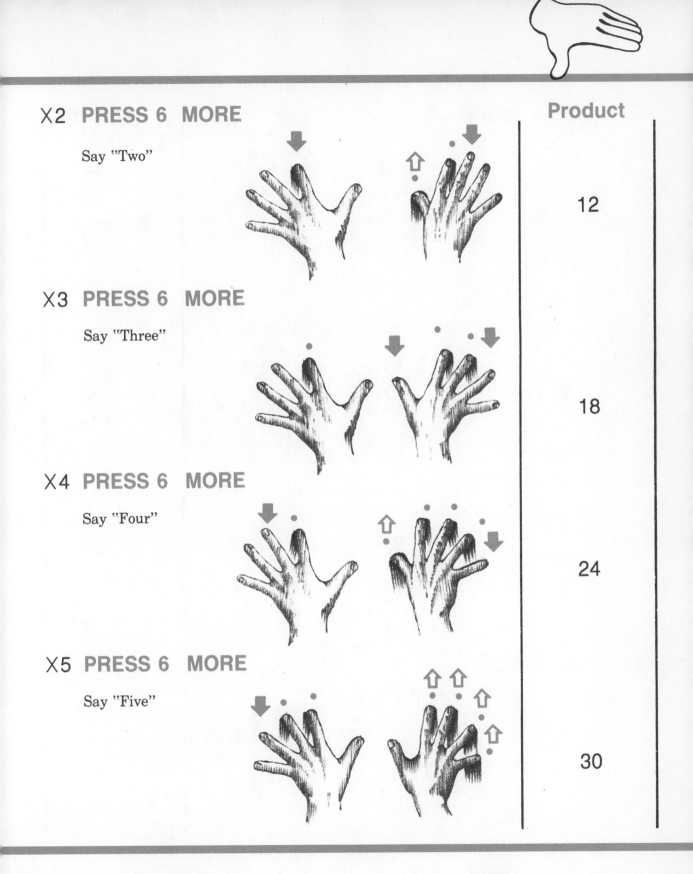

6 S

X6 **PRESS 6 MORE**

Say "Six"

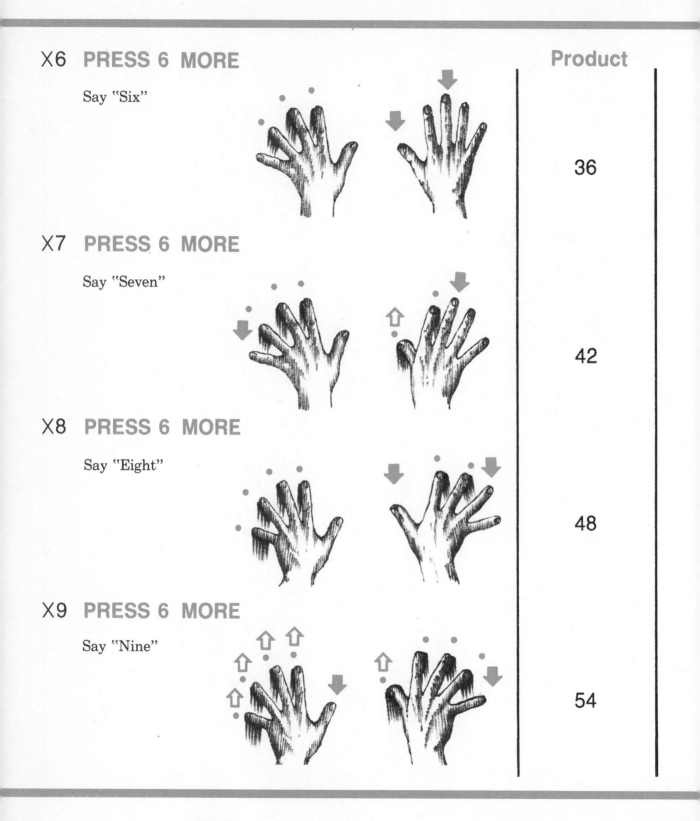

Product

36

X7 **PRESS 6 MORE**

Say "Seven"

42

X8 **PRESS 6 MORE**

Say "Eight"

48

X9 **PRESS 6 MORE**

Say "Nine"

54

224

X10 PRESS 6 MORE

Say "Ten"

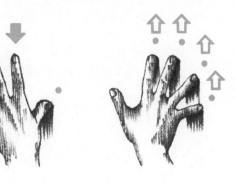

Product

60

Press this sequence again from beginning to end, counting the instant presses until you have counted off "ten." Repeat it with your eyes closed.

Now for a quick review, to freshen up all you have learned so far about multiplication. Go back to page 190 and repeat the pressing of 2s ten times (10 × 2). Then turn to page 204 and repeat the pressing of 5s ten times (10 × 5). Finally, repeat the pressing of 6s ten times (10 × 6), as shown above.

Then do the examples on the following pages. They'll give you plenty of practice with everything you've learned. Don't cheat yourself; press *every* example. When you're finished, have someone call them out to you. Then repeat them with your eyes closed.

Here is a series of exercises that will reinforce your ability to press 6s repeatedly when they do not occur in the now familiar sequence which results from an initial press of 6. You still will use one of the established ways of pressing each 6.

First, Press 1. Then press 6s ten times.
When finished, your fingers read 61.
Next, Press 2. Then press 6s ten times.
When finished, your fingers read 62.
Next, Press 3. Then press 6s ten times.
When finished, your fingers read 63.

Continue this pattern, first pressing each of the numbers up to 9, followed by ten consecutive 6s. Note that at the end of each series of ten

6s you easily can recognize the correct result because the number first pressed clearly is part of the final sum.

Take your time and whenever you falter on any set of manipulations, go over it again until you're sure of it.

Do not proceed to the next number, 9, until you feel strong with 2s, 5s and 6s.

Multiplying 2s, 5s and 6s

This entire page to be repeated orally.

$2 \times 2 =$ $2 \times 6 =$ $7 \times 5 =$	$4 \times 2 =$ $4 \times 6 =$ $9 \times 5 =$	$1 \times 5 =$ $6 \times 2 =$ $6 \times 6 =$	$3 \times 5 =$ $8 \times 2 =$ $8 \times 6 =$	$5 \times 5 =$ $10 \times 2 =$ $10 \times 6 =$
$1 \times 5 =$ $3 \times 6 =$ $3 \times 2 =$	$3 \times 5 =$ $5 \times 5 =$ $1 \times 2 =$	$10 \times 2 =$ $6 \times 5 =$ $7 \times 6 =$	$7 \times 2 =$ $8 \times 5 =$ $8 \times 6 =$	$1 \times 6 =$ $5 \times 2 =$ $9 \times 6 =$
$10 \times 6 =$ $14 \times 5 =$ $17 \times 2 =$	$10 \times 5 =$ $0 \times 2 =$ $11 \times 6 =$	$11 \times 5 =$ $20 \times 2 =$ $12 \times 6 =$	$12 \times 5 =$ $19 \times 2 =$ $13 \times 6 =$	$13 \times 5 =$ $18 \times 2 =$ $14 \times 6 =$

Pressing 2s, 5s and 6s - Direct and with Exchanges* - Addition

6 5 2 + 6	5 6 6 + 6	6 2 5 + 6	6 6 5 2 + 6●	8 5 6 6● + 2	9 6● 5 6 + 5	5 6 2 5 + 6	6 5 5 2 + 5

*Exchanges marked for 6s using the 10–4 manipulation show as ●

6 S

227

9s

In the midst of learning something like Fingermath, where the particulars of technique are so numerous, you are apt to think you are standing still because progress is so gradual. Yet your development must surely be more perceptible now that you have gone through these last few pages of review. Had you realized how far you had come? My guess is that you hadn't. If you make it a point to spend as little as ten minutes a day doing written and oral reviews, you'll find before long that you possess new skills that will serve you all your life.

Having labored through a succession of 2s, 5s, and 6s, using them both to add and to multiply, you have come to realize that the ability to choose the right manipulation is important in meeting every arithmetic challenge. You also know, through your own experience, that making the right choice becomes less of a conscious selection, more automatic, the more you practice.

There remain five numbers (9, 8, 4, 7, 3) that you must learn to multiply or otherwise manipulate in a direct manner. With each of them, let's first explore all the possible ways of making a single, instant press.

The number 9 offers another relatively simple sequence of manipulations. There are only two ways of registering 9.

Two Ways To Press 9

9 (full Right Hand press)

10–1 (Left Hand press 10, Right Hand clear 1)

Go through these 9s the first several times by following only the diagrams, ignoring the oral count and the columns of notations to the left and right. Concentrate *only* on observing, feeling, and comprehending the manipulations themselves.

⑨ PRESS 9

×1   say "One"

9

×2 **PRESS 9 MORE**

say "Two"

18

(You are simultaneously pressing 10 and clearing 1. Remember the reverse clearing rule: Last in, first out.)

×3 **PRESS 9 MORE**

say "Three"

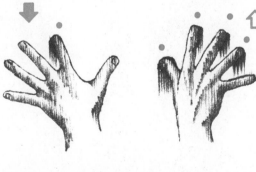

27

(Another press 10, clear 1)

✕4 **PRESS 9 MORE**

say "Four"

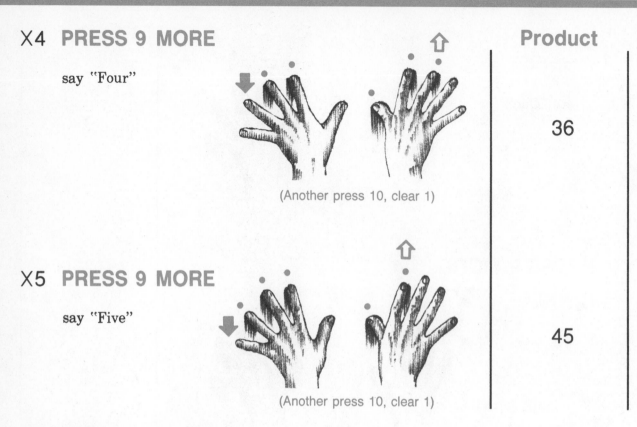

(Another press 10, clear 1)

✕5 **PRESS 9 MORE**

say "Five"

(Another press 10, clear 1)

	Product
	36
	45

✕6 **PRESS 9 MORE**

The next 9 follows the same pattern of pressing 10 and clearing 1. But it seems more involved until you learn a simple trick for handling it. First, think your way through it this way:

W A I T
(Hold onto 45)

45

You've got 45. To press 10, you perform the 40 to 50 exchange on your left hand. To clear 1, you perform the 5 to 4 exchange on your right. Forward on your left and reverse on your right. Like this:

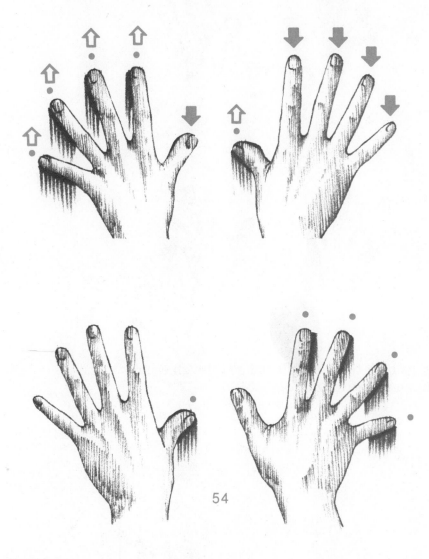

What you're doing, of course, is pressing all the fingers that had been clear and clearing all the fingers that had been pressed. It's a simple flip-flop. Do it again. Press 45. Now flip up the pressed fingers as you press down the clear fingers. Try this in reverse: go from 54 to 45, then back again.

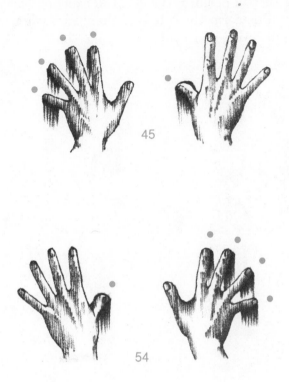

45

54

Now, clear all your fingers, go back to the first 9 and go through each press to 45. Then see if you can maneuver right into 54.

Now continue . . .

Product

×7 **PRESS 9 MORE**

    say "Seven"

(10 − 1)

63

✗8 PRESS 9 MORE

say "Eight"

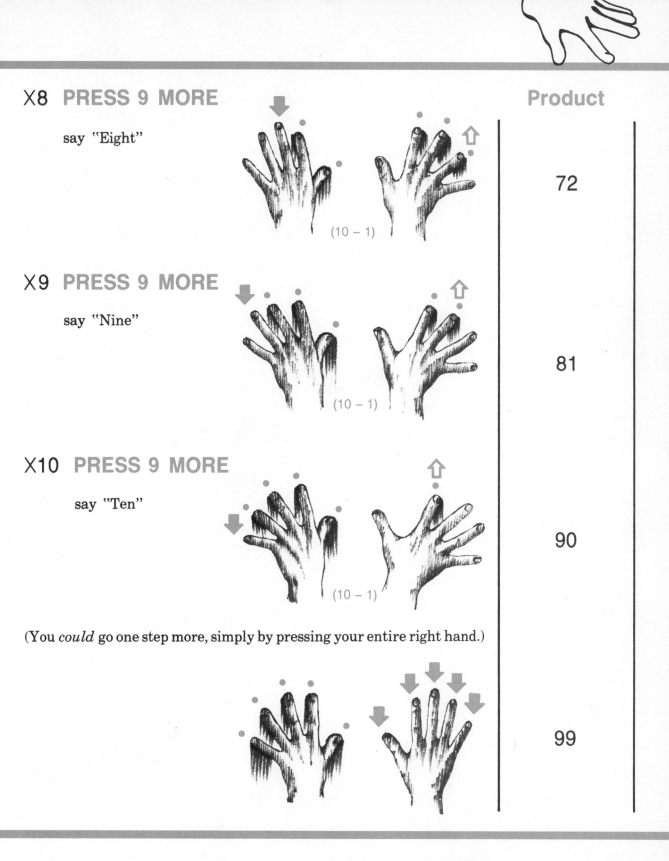

(10 – 1)

Product

72

✗9 PRESS 9 MORE

say "Nine"

(10 – 1)

81

✗10 PRESS 9 MORE

say "Ten"

(10 – 1)

90

(You *could* go one step more, simply by pressing your entire right hand.)

99

Return to the beginning of this 9s sequence and go through it again many times, still ignoring the accumulated partial products as you proceed and concentrating only on crisp, steady manipulations. Work at the 45 to 54 exchange until it falls into place without hesitation.

Then begin again, this time (as I've noted for you) counting the number of times you press 9 until you reach 99.

Here is a series of exercises that will reinforce your ability to press 9s repeatedly when they do not occur in the now familiar sequence which results from an initial press of 9. You still will use one of the established ways of pressing each 9.

First, Press 1. Then press 9s ten times.
When finished, your fingers read 91.
Next, Press 2. Then press 9s ten times.
When finished, your fingers read 92.
Next, Press 3. Then press 9s ten times.
When finished, your fingers read 93.

Continue this pattern, first pressing each of the numbers up to 9, followed by ten consecutive 9s. Note that at the end of each series of ten 9s you easily can recognize the correct result because the number first pressed clearly is part of the final sum.

When you feel secure with the complete run of 9s—and *only* when you feel secure—try your hands at the exercises that follow. You should be able to navigate at a steady pace from 9 to 99 without pausing, and counting the number of times you press. Music is still the best pacer you can use because you'll always know when you're missing the beat.

Multiplying 2s, 5s, 6s and 9s

This entire page to be repeated orally.

$6 \times 9 =$ $7 \times 6 =$ $8 \times 5 =$	$7 \times 9 =$ $8 \times 6 =$ $9 \times 5 =$	$8 \times 9 =$ $9 \times 6 =$ $10 \times 5 =$	$9 \times 9 =$ $10 \times 6 =$ $11 \times 5 =$	$10 \times 9 =$ $11 \times 6 =$ $12 \times 5 =$
$9 \times 2 =$ $12 \times 6 =$ $10 \times 6 =$	$10 \times 2 =$ $6 \times 6 =$ $1 \times 5 =$	$11 \times 2 =$ $7 \times 6 =$ $3 \times 5 =$	$12 \times 2 =$ $8 \times 6 =$ $5 \times 5 =$	$13 \times 2 =$ $9 \times 6 =$ $7 \times 5 =$
$9 \times 5 =$ $11 \times 9 =$ $5 \times 2 =$	$3 \times 9 =$ $6 \times 2 =$ $0 \times 6 =$	$5 \times 9 =$ $8 \times 2 =$ $3 \times 6 =$	$7 \times 9 =$ $10 \times 2 =$ $5 \times 6 =$	$9 \times 9 =$ $3 \times 2 =$ $7 \times 6 =$

Pressing 2s, 5s, 6s, and 9s - Direct and with Exchanges

						6	2
9	5	9	6	2	5	2	9
5	9	9	5	9	6	6	9
9	6	2	9	2	6	6	6
9	9	6	5	9	9	9	9
$+9$	$+2$	$+9$	$+9$	$+2$	$+9$	$+9$	$+9$

8s

You're getting into some very sophisticated moves now. As I've pointed out to you several times, once you learn to press whole numbers, responding confidently to written or oral instructions, you will be able to use these manipulations for each of the four basic operations of arithmetic—addition, subtraction, multiplication, and division—no matter how challenging a particular exercise may seem.

You have already learned 1s, 2s, 5s, 6s, and 9s. Only four others remain: 8s, 4s, 7s, and 3s. It's time to tackle 8s.

Because you're gaining an appreciation for the logic of Fingermath manipulations, I don't have to continue to take time to go into as much detail as I did at the outset. I'll show you the pressing possibilities for the number 8, and then provide you with a complete series of diagrams.

Three Ways to Press 8

5 + 3 (Right Hand)

5 Exchange + 3

10 − 2

Before you begin, remember to go through the entire sequence first, ignoring the oral count and the notations to the left and right of the page. Concentrate only on executing the manipulations crisply and steadily. Once you've acquired that skill, go back and count off the number of times you press 8, as you would for multiplication.

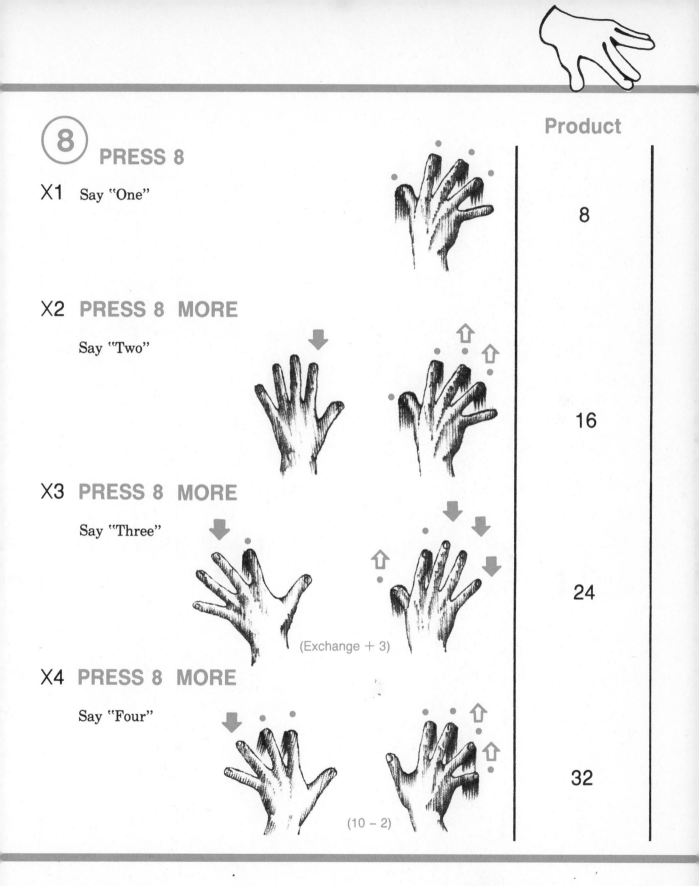

(8) PRESS 8

Product

X1 Say "One"

8

X2 **PRESS 8 MORE**

Say "Two"

16

X3 **PRESS 8 MORE**

Say "Three"

(Exchange + 3)

24

X4 **PRESS 8 MORE**

Say "Four"

(10 − 2)

32

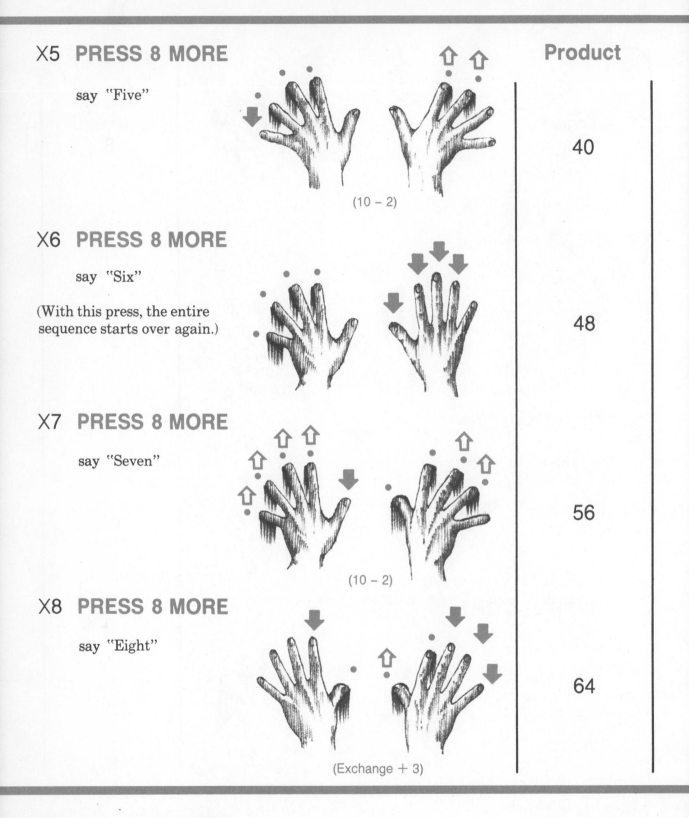

×5 **PRESS 8 MORE**

say "Five"

(10 − 2)

Product

40

×6 **PRESS 8 MORE**

say "Six"

(With this press, the entire
sequence starts over again.)

48

×7 **PRESS 8 MORE**

say "Seven"

(10 − 2)

56

×8 **PRESS 8 MORE**

say "Eight"

(Exchange + 3)

64

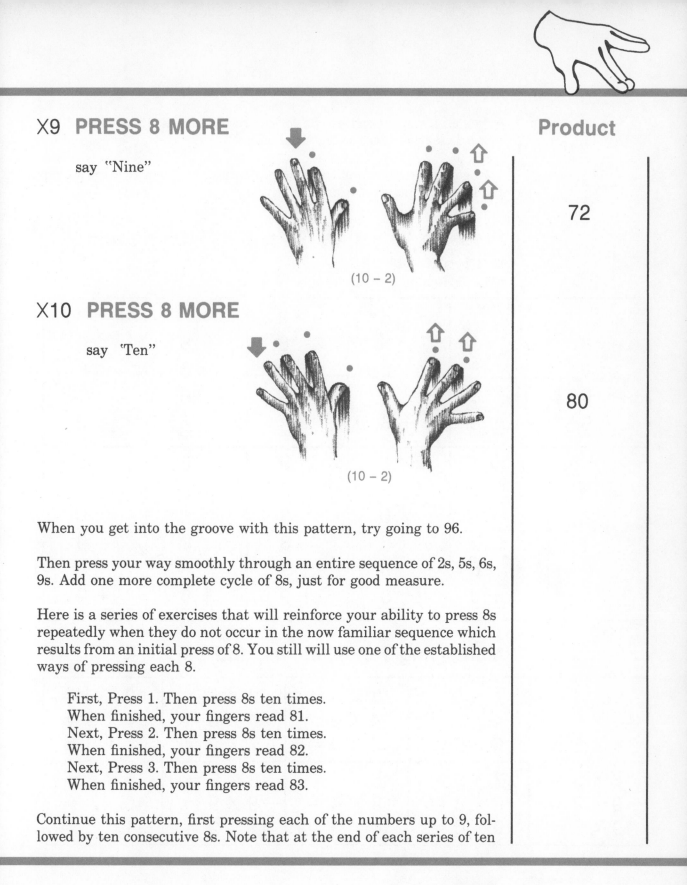

X9 **PRESS 8 MORE**

Product

say "Nine"

(10 – 2)

72

X10 **PRESS 8 MORE**

say "Ten"

(10 – 2)

80

When you get into the groove with this pattern, try going to 96.

Then press your way smoothly through an entire sequence of 2s, 5s, 6s, 9s. Add one more complete cycle of 8s, just for good measure.

Here is a series of exercises that will reinforce your ability to press 8s repeatedly when they do not occur in the now familiar sequence which results from an initial press of 8. You still will use one of the established ways of pressing each 8.

First, Press 1. Then press 8s ten times.
When finished, your fingers read 81.
Next, Press 2. Then press 8s ten times.
When finished, your fingers read 82.
Next, Press 3. Then press 8s ten times.
When finished, your fingers read 83.

Continue this pattern, first pressing each of the numbers up to 9, followed by ten consecutive 8s. Note that at the end of each series of ten

8s you easily can recognize the correct result because the number first pressed clearly is part of the final sum.

Now go on and put all of these manipulations together by working on the following pages of examples.

Pressing 2s, 5s, 6s, 9s and 8s - Direct and with Exchanges
This page to be practiced orally.

		5	9	8	9
8	2	8	8	8	8
9	8	6	2	6	8
6	8	8	6	6	8
+ 5	+ 8	+ 8	+ 8	+ 8	+ 2

			9	8	5
8	5	8	8	8	6
2	5	5	8	9	8
8	6	8	8	9	9
2	8	5	9	6	8
+ 8	+ 8	+ 8	+ 9	+ 6	+ 9

Multiplying 2s, 5s, 6s, 9s, and 8s

6 X 8 =	7 X 8 =	8 X 8 =	9 X 8 =	10 X 8 =
3 X 5 =	4 X 2 =	4 X 9 =	4 X 6 =	4 X 5 =
6 X 6 =	6 X 5 =	6 X 9 =	7 X 2 =	7 X 6 =
7 X 5 =	7 X 9 =	8 X 2 =	8 X 5 =	8 X 6 =
9 X 2 =	9 X 8 =	9 X 6 =	9 X 9 =	10 X 8 =
10 X 5 =	10 X 6 =	10 X 9 =	11 X 2 =	11 X 5 =

Review: Addition and Multiplication Facts

69,582 + 74,865	$12 \times 8 = \square$	894,652 35,769 + 40,828	$27 + \square = 46$
$7 \times 9 = \square$	37,489 + 68,688	$15 \times 5 = \square$	129,568 358,956 269,625 + 456,890
$11 \times 6 = \square$	$X + 8 + 18 = 34$ $X = \square$	$23 \times 2 = \square$	$54 + 5 + \square = 60$
$33 + Y = 51$ $Y = \square$	$\square + 12 + 8 = 31$	56,982 4,068 37,259 + 3,646	$13 + 6 + \square = 38$
$4 \times 8 + 6 = \square$	85,953 93,868 25,935 88,888 + 27,436	$27 + 9 + Y = 47$ $Y = \square$	$11 \times 9 = \square$

4s

You're coming into the final turn. Most of the needed multiplication facts are in your head and fingers, and you have most of the insight and skill you'll need for the remaining numbers, the 4s, 7s, and 3s.

Approach the 4s in the same manner that you have used with all previous numbers. First work your way through the mechanical sequence of 4s. When you can execute these steadily and with confidence, then proceed through the sequence with oral counting. Do that at least ten times. If music has helped you with other numbers, use it again to achieve regularity.

Four Ways to Press 4

4 (Right Hand)

5 – 1 (Right Hand)

5 Exchange – 1

10 – 6

 PRESS 4

X1   Say "One"

Product

4

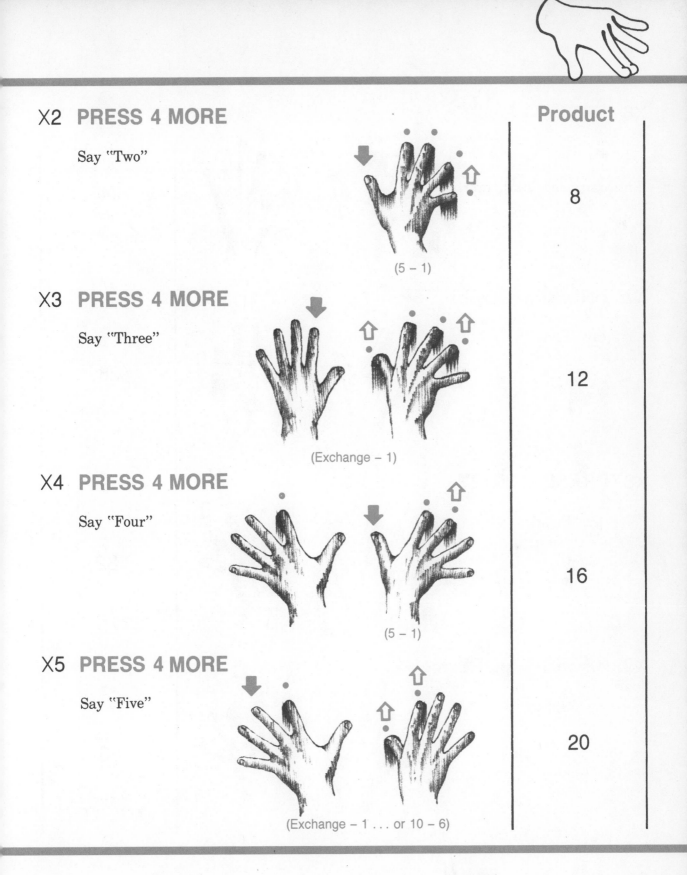

		Product
X2 **PRESS 4 MORE** Say "Two"	(5 – 1)	8
X3 **PRESS 4 MORE** Say "Three"	(Exchange – 1)	12
X4 **PRESS 4 MORE** Say "Four"	(5 – 1)	16
X5 **PRESS 4 MORE** Say "Five"	(Exchange – 1 . . . or 10 – 6)	20

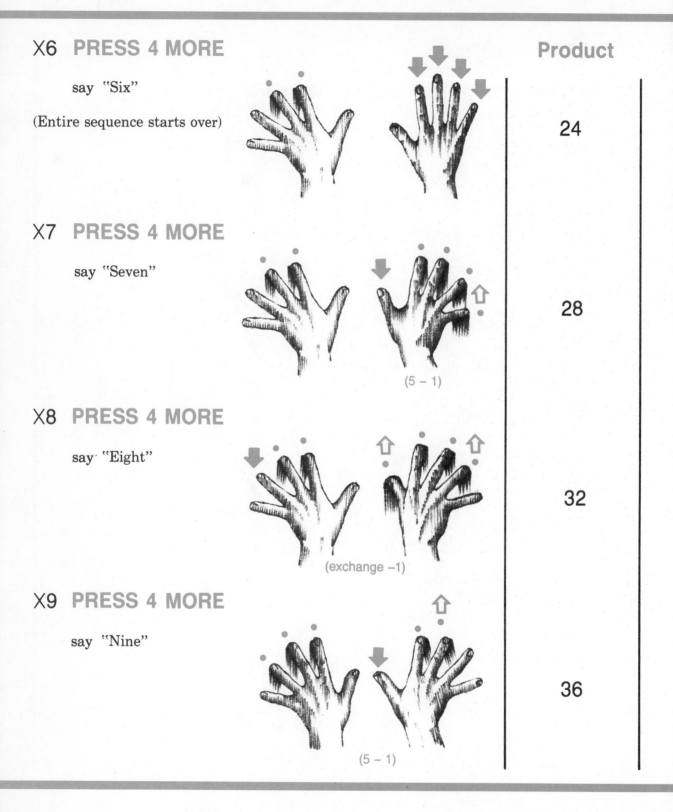

X6 **PRESS 4 MORE**   Product

say "Six"

(Entire sequence starts over)

24

X7 **PRESS 4 MORE**

say "Seven"

28

(5 − 1)

X8 **PRESS 4 MORE**

say "Eight"

32

(exchange −1)

X9 **PRESS 4 MORE**

say "Nine"

36

(5 − 1)

X10 PRESS 4 MORE

say "Ten"

(exchange −1 . . . or 10 −6)

Product

40

Try to go beyond a product of 40. You can go as high as 96. (Later, when you learn how to pass 99, the sky will be the limit!)

Here is a series of exercises that will reinforce your ability to press 4s repeatedly when they do not occur in the now familiar sequence which results from an initial press of 4. You still will use one of the established ways of pressing each 4.

First, Press 1. Then press 4s ten times.
When finished, your fingers read 41.
Next, Press 2. Then press 4s ten times.
When finished, your fingers read 42.
Next, Press 3. Then press 4s ten times.
When finished, your fingers read 43.

Continue this pattern, first pressing each of the numbers up to 9, followed by ten consecutive 4s. Note that at the end of each series of ten 4s you easily can recognize the correct result because the number first pressed clearly is part of the final sum.

4 S

Multiplication of 2s, 5s, 6s, 9s, 8s and 4s
This entire page to be repeated orally.

$5 \times 8 =$	$6 \times 8 =$	$6 \times 4 =$	$7 \times 4 =$	$8 \times 4 =$
$9 \times 4 =$	$10 \times 4 =$	$11 \times 4 =$	$14 \times 2 =$	$13 \times 2 =$
$12 \times 2 =$	$11 \times 2 =$	$10 \times 2 =$	$9 \times 2 =$	$7 \times 8 =$
$8 \times 8 =$	$9 \times 8 =$	$10 \times 8 =$	$11 \times 8 =$	$12 \times 8 =$
$12 \times 4 =$	$0 \times 4 =$	$4 \times 2 =$	$3 \times 2 =$	$15 \times 5 =$
$14 \times 5 =$	$13 \times 5 =$	$12 \times 5 =$	$11 \times 5 =$	$10 \times 5 =$
$10 \times 4 =$	$12 \times 4 =$	$1 \times 4 =$	$3 \times 4 =$	$5 \times 4 =$
$7 \times 4 =$	$2 \times 2 =$	$1 \times 2 =$	$0 \times 2 =$	$15 \times 6 =$
$14 \times 6 =$	$13 \times 6 =$	$9 \times 5 =$	$8 \times 5 =$	$7 \times 5 =$

Pressing 2s, 5s, 6s, 9s, 8s and 4s - Direct and with Exchanges

6 4 6 + 4	4 4 5 + 4	2 4 2 + 4	4 4 4 + 4	9 4 5 + 4	8 4 6 4 + 4	9 4 5 2 + 4
4 6 4 9 + 4	2 5 4 5 + 4	9 6 4 5 4 + 4	4 8 8 5 6 + 6	5 4 8 4 9 + 9	8 4 4 4 4 + 5	2 4 6 8 9 + 5

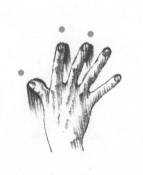

7s

The home stretch. Only 7s and 3s are still to come, and then you'll have them all. Even though I've told you that these numbers for multiplication are being presented in the order of increasing difficulty, that's only a relative term. Not one of these number facts is a real problem child, simply because each is so logically structured and easily understood. You are never confronted with a sudden turn into unfamiliar territory. To the contrary, each set of manipulations prepares you better for those that follow. So stretch your fingers, crack your knuckles, ponder the possibilities, and dive into 7s.

Three Ways To Press 7

5 + 2

5 Exchange + 2

10 − 3

 **PRESS 7**

X1   say "One"

Product

7

X2 PRESS 7 MORE

say "Two"

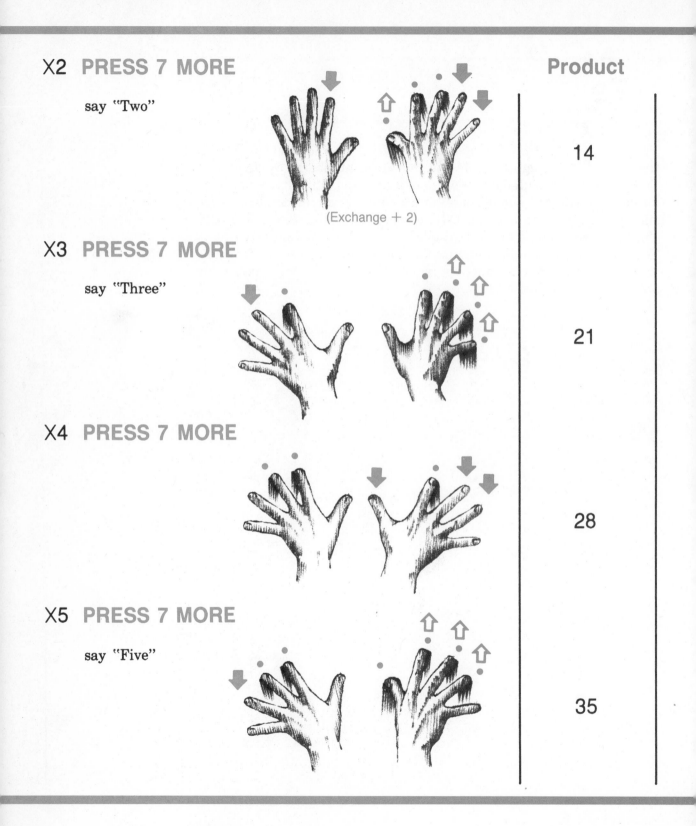

(Exchange + 2)

Product

14

X3 PRESS 7 MORE

say "Three"

21

X4 PRESS 7 MORE

28

X5 PRESS 7 MORE

say "Five"

35

248

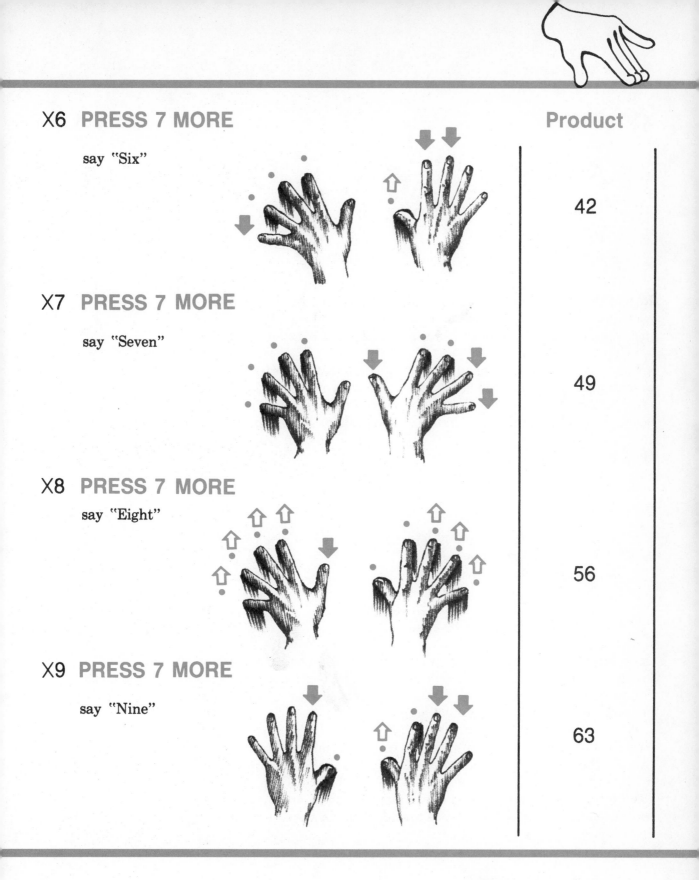

X6 **PRESS 7 MORE**

say "Six"

Product

42

X7 **PRESS 7 MORE**

say "Seven"

49

X8 **PRESS 7 MORE**

say "Eight"

56

X9 **PRESS 7 MORE**

say "Nine"

63

X10 PRESS 7 more

say "Ten"

70

Repeat the entire sequence, continuing to observe finger positions after each new press of 7. You must practice this until each press is an instantaneous whole number 7, not a combination of separate parts.

When you are able to maneuver evenly through this series without breaking your rhythm, you are ready to use it for multiplication of 7s. Start again, this time counting the number of times you press 7. Go beyond a product of 70. Your hands are capable of calculating to 98.

Here is a series of exercises that will reinforce your ability to press 7s repeatedly when they do not occur in the now familiar sequence which results from an initial press of 7. You still will use one of the established ways of pressing each 7.

> First, Press 1. Then press 7s ten times.
> When finished, your fingers read 71.
> Next, Press 2. Then press 7s ten times.
> When finished, your fingers read 72.
> Next, Press 3. Then press 7s ten times.
> When finished, your fingers read 73.

Continue this pattern, first pressing each of the numbers up to 9, followed by ten consecutive 7s. Note that at the end of each series of ten 7s you easily can recognize the correct result because the number first pressed clearly is part of the final sum.

Now review every one of the multipliers you have learned. First execute them in the order I gave them to you (2, 5, 6, 9, 8, 4, 7). Then mix them up. Before performing each one, review the possible ways of pressing it. Then attempt to count through the sequence, slowly, at an even pace, without stopping, until both hands are filled to capacity.

Multiplication of 2s, 5s, 6s, 9s, 8s, 4s and 7s

This entire page to be repeated orally.

$2 \times 7 =$	$4 \times 7 =$	$6 \times 7 =$	$8 \times 2 =$	$4 \times 2 =$
$7 \times 2 =$	$3 \times 4 =$	$7 \times 4 =$	$11 \times 4 =$	$8 \times 7 =$
$10 \times 7 =$	$8 \times 7 =$	$4 \times 5 =$	$8 \times 5 =$	$7 \times 5 =$
$9 \times 8 =$	$5 \times 8 =$	$1 \times 8 =$	$7 \times 7 =$	$5 \times 7 =$
$3 \times 7 =$	$8 \times 6 =$	$4 \times 6 =$	$0 \times 6 =$	$5 \times 9 =$
$9 \times 9 =$	$1 \times 9 =$	$1 \times 7 =$	$11 \times 7 =$	$13 \times 7 =$
$3 \times 9 =$	$7 \times 9 =$	$8 \times 9 =$	$11 \times 6 =$	$7 \times 6 =$
$3 \times 6 =$	$2 \times 7 =$	$6 \times 7 =$	$10 \times 7 =$	$8 \times 8 =$
$4 \times 8 =$	$0 \times 8 =$	$9 \times 5 =$	$5 \times 5 =$	$1 \times 5 =$

Pressing 2s, 5s, 6s, 9s, 8s, 4s and 7s - Direct and with Exchanges

						8
7	4	9	5	7	7	7
4	7	7	7	6	9	7
7	6	7	5	5	9	9
+7	+7	+7	+7	+7	+7	+7
					5	4
9	7	5	9	4	6	7
6	5	7	2	5	7	7
2	7	5	7	7	7	4
7	7	7	1	4	8	6
+7	+6	+6	+7	+7	+7	+4

3s

Though you may not have been aware of it, most of the sequences you have been following have been helping to prepare you for Fingermath subtraction. Here are just a few of the equivalents you have learned:

$$2 = 10 - 8$$
$$2 = 5 - 3$$
$$5 = 10 - 5$$
$$6 = 10 - 4$$
$$9 = 10 - 1$$
$$8 = 10 - 2$$
$$4 = 10 - 6$$
$$4 = 5 - 1$$
$$7 = 10 - 3$$

Each of these utilizes a simultaneous press and clear combination. With practice, they will become automatic, and in time, make your subtraction tasks relatively simple.

Another such set occurs in this last series of multiplication facts, dealing with the number 3.

Three Ways to Press 3

3 (Right Hand)

5 – 2 (Exchange)

5 Exchange – 2 *or*

10 – 7 (Same as 5 Exchange – 2)

As before, first concentrate solely on working out each press of 3 according to the diagrams. You will need more practice on this series than on previous ones. So take your time and learn it well. You will not feel secure or comfortable with 3s or any other number until you have practiced each frequently, so don't get the feeling that you're all thumbs! Fingermath is like any other activity that requires motor skills and coordination. Practice and use will make you an expert. Count on it.

PRESS 3

X1   say "One"

X2 **PRESS 3 MORE**

say "Two"

X3 **PRESS 3 MORE**

say "Three"

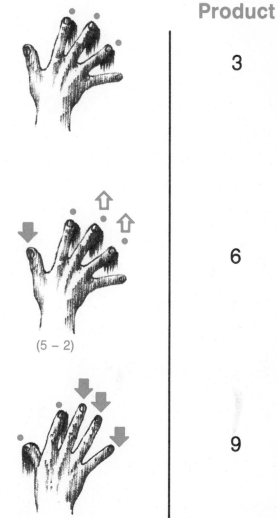

(5 – 2)

Product

3

6

9

X4 **PRESS 3 MORE**   Product

say "Four"

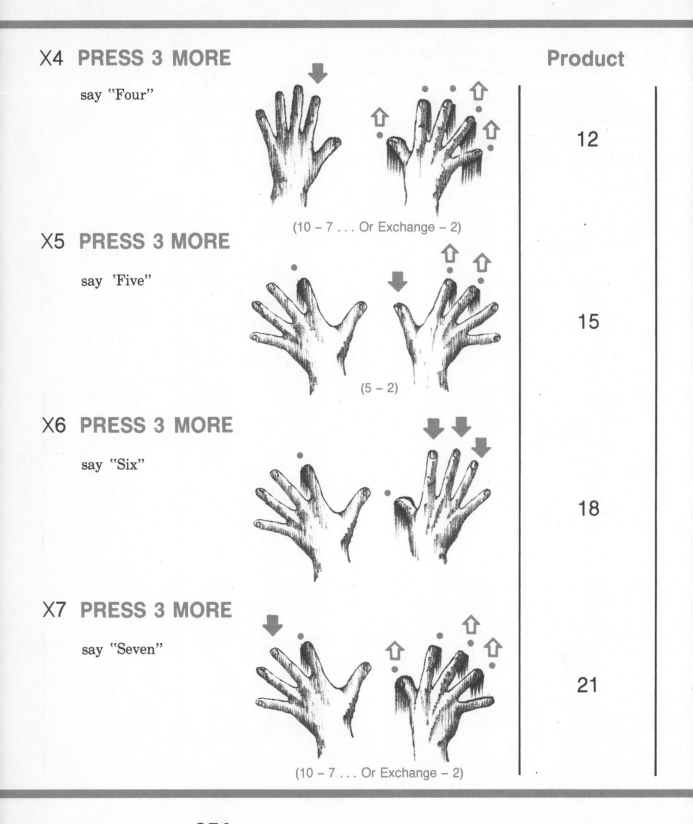

(10 – 7 . . . Or Exchange – 2)   12

X5 **PRESS 3 MORE**

say 'Five"

(5 – 2)   15

X6 **PRESS 3 MORE**

say "Six"

                                                         18

X7 **PRESS 3 MORE**

say "Seven"

                                                         21

(10 – 7 . . . Or Exchange – 2)

MULTIPLICATION

X8 PRESS 3 MORE

Product

say "Eight"

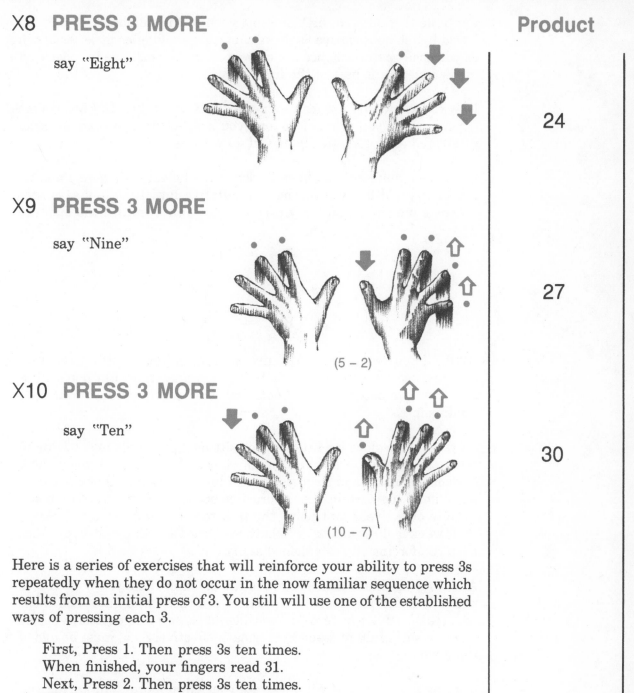

24

X9 PRESS 3 MORE

say "Nine"

(5 – 2)

27

X10 PRESS 3 MORE

say "Ten"

(10 – 7)

30

Here is a series of exercises that will reinforce your ability to press 3s repeatedly when they do not occur in the now familiar sequence which results from an initial press of 3. You still will use one of the established ways of pressing each 3.

First, Press 1. Then press 3s ten times.
When finished, your fingers read 31.
Next, Press 2. Then press 3s ten times.
When finished, your fingers read 32.
Next, Press 4. Then press 3s ten times.
When finished, your fingers read 34.

3 S

Continue this pattern, first pressing each of the numbers up to 9, followed by ten consecutive 3s. Note that at the end of each series of ten 3s you easily can recognize the correct result because the number first pressed clearly is part of the final sum.

Now you know why I saved 3s for last! They are dillies. But believe me when I assure you that before long you'll whip through them as accurately and as automatically as you do with 5s.

Until then, however, you have an ace in the hole. If you were a mathematician you'd know it as the commutative law. It simply says that numbers multiplied one way produce the same result in the opposite way.

$$6 \times 7 \quad = \quad 7 \times 6$$

$$9 \times 3 \quad = \quad 3 \times 9$$

$$A \times B \quad = \quad B \times A$$

So, until you really become secure with all of your multiplication sequences, use the ones you know best in reverse order. But don't allow this crutch to become permanent. You must continue to work at your weak points.

Work on them slowly; a little at each sitting. Spend only about 10 or 15 minutes at a time practicing any of your skills. But go through these sessions regularly and frequently . . . at least once a day. If you're teaching a child, be certain that a practice schedule is set up that has a definite objective. Don't leave children to their own devices. Observe their weaker skills and be certain these are given a daily workout. Also, keep reinforcing the established skills.

And since you know the rules of the game, follow them yourself. You've probably heard the expression, "There are those who know and those who teach." But I have never considered these to be exclusive of one another. Certainly to teach Fingermath effectively you must be one of those who know!

Nothing I can think of will motivate your child more than having you participate and even compete in the project. You'd better be on your toes, though, because I haven't yet met the child who couldn't outperform his elders in Fingermath. Ah, youth!

Pressing 2s, 5s, 6s, 9s, 8s, 4s, 7s, 3s -
All Direct and with Exchanges

This entire page to be repeated orally.

3	3	7	7	3	7	4	3
3	7	7	3	7	3	3	4
7	3	3	7	7	3	7	3
+7	+7	+3	+3	+3	+7	+3	+7

3	9	3	6	3	7	6	8
9	7	8	3	7	6	3	8
7	3	9	7	3	9	3	7
9	3	9	8	6	8	3	7
+3	+7	+7	+9	+8	+3	+7	+3

Multiplication of 2s, 5s, 6s, 9s, 8s, 4s, 7s and 3s

$1 \times 3 =$	$2 \times 3 =$	$3 \times 3 =$	$4 \times 3 =$	$5 \times 3 =$
$6 \times 3 =$	$2 \times 2 =$	$3 \times 4 =$	$4 \times 5 =$	$5 \times 6 =$
$6 \times 7 =$	$7 \times 8 =$	$11 \times 8 =$	$10 \times 7 =$	$9 \times 6 =$
$8 \times 5 =$	$7 \times 4 =$	$6 \times 2 =$	$7 \times 3 =$	$8 \times 3 =$
$9 \times 3 =$	$10 \times 3 =$	$11 \times 3 =$	$12 \times 3 =$	$8 \times 9 =$
$9 \times 1 =$	$3 \times 2 =$	$5 \times 4 =$	$6 \times 5 =$	$7 \times 6 =$

Review: All Skills With All Numbers

950,673 + 921,897	$13 + \square = 27$	$11 \times 3 =$	$12 \times 4 =$
$15 \times 5 =$	$9 \times 8 =$	$Y + 37 = 46$ $Y = \square$	8,967,431 + 4,987,339
$14 \times 6 =$	$18 \times 2 =$	$8 \times 3 + 9 = \square$	928,637 − 48,888
9 −3 4 −5	11 $\times$ 6	63,784 80,695 42,532 8,704 + 671	$21 \times 2 - 6 = \square$
$16 \times 3 =$	4,867,210 3,754,658 + 2,945,762	12 $\times$ 8	368,955 −219,666
$12 \times 4 - \square = 29$	49 $\times$ 2	$Y + 8 = 46$ $Y = \square$	$23 \times 4 =$
9 $\times$7	$54 - \square = 45$	294,167 − 95,638	8 $\times$8

Divide And Conquer Your Fears

Throughout this book, I have emphasized the fact that addition is the most essential operation of arithmetic. The other operations are based logically on addition, as you have already seen with multiplication (repeated addition) and subtraction (reversed addition).

Division is just as subordinate to addition. Yet this simple operation has the reputation of being an unpredictable exercise that requires more guesswork than reasoning.

Not so. We must only approach division with some patience and understanding to appreciate its fundamental structure.

$$20 \div 5 = \square$$

What do these symbols mean? They can be translated into ordinary language in several ways. For example: The number 20 divided by 5 equals how many? Or: If one divides the number 20 into five parts, how many will be contained in each part?

You may also see the same problem set forth in several symbolic ways:

$$20 \div 5 = \square \qquad 5 \overline{)\,20} \qquad \frac{20}{5} = \square$$

Symbols and words, they all ask the same question: How many times is the number 5 contained in the number 20?

Achieving the answer through Fingermath is so simple that you've probably thought through the procedure already. But I'll outline it quickly anyway just in case you haven't. Just as you previously employed addition to multiply, so will you use the same skills to divide.

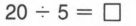

$$20 \div 5 = \square$$

In this exercise, all you need to do is press 5s, counting the number of times you are pressing, until you recognize 20.

PRESS 5

Say "One"

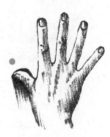

PRESS 5 MORE

Say "Two"

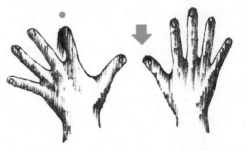

PRESS 5 MORE

Say "Three"

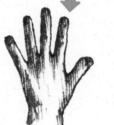

PRESS 5 MORE

Say "Four"

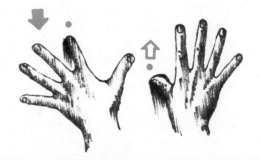

There it is—20!

You counted to 4

THEREFORE $20 \div 5 = 4$

If your answer is correct, you have determined that four 5s are contained in 20. To prove your calculation, add up four 5s, or five 4s, and see if you end up with 20.

In the example above, you have divided a larger number by a smaller one ($20 \div 5$). In multiplication, as you remember, the order of the numbers makes no difference: 20×5, for example, has the same value as 5×20. This rule does not hold in division. If you switch the order of the numbers, and try to divide the smaller number by the larger ($5 \div 20$), you find yourself in the realm of fractions.

A word about fractions is in order at this point. As their name suggests, they are the result of breaking up (fracturing) a whole into parts. This is not a counting procedure, and it cannot be simply dealt with by using the techniques of addition. Further, even in the addition of fractions, the simple Fingermath techniques do not apply. Fingermath is a method that is confined to the basic operations of adding, subtracting, multiplying, and dividing whole numbers.

In the division examples that follow, we shall ignore fractions and deal only with problems in which the divisor (the number following the division sign) is smaller than the number to be divided.

Unlike the first example, not all division problems will come out even. Even or not, this should not complicate matters if you use Fingermath to get the answer. Since division is based on multiplication, one very effective way to employ Fingermath is to work with basic multiplication facts, in exactly the same way that is familiar. Consider this example:

$$8 \overline{)39}$$

Without Fingermath, you would have to know your 8s table to figure out this one. You would estimate the number of 8s in 39 and then work out the remainder.

With Fingermath, there's no need to have memorized anything. If you can perform your 8s manipulations in sequence, you're home free!

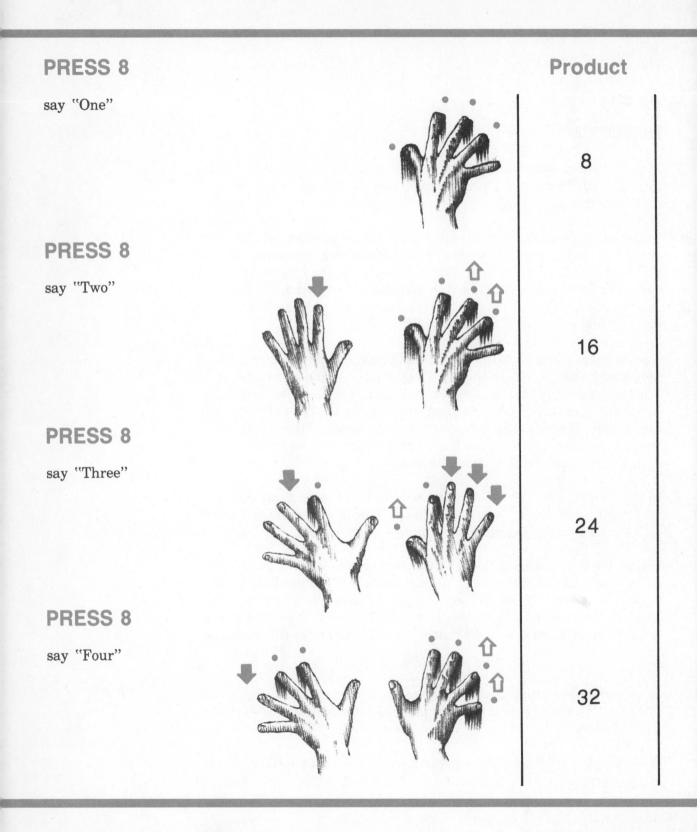

	Product
PRESS 8 say "One"	8
PRESS 8 say "Two"	16
PRESS 8 say "Three"	24
PRESS 8 say "Four"	32

If you press the next 8, you'll find that you've passed the number you want to reach, which is 39. So always stop at the press that takes you almost up to but never beyond the dividend (the number being divided).

You are now at 32.
You have pressed 8 four times.
Your answer *so far* is 4.
Enter 4 as your partial answer and show the 32 below the 39, like this:

$$8\overline{)39} \quad \begin{array}{r} 4\ r7 \\ \underline{32} \\ 7 \end{array}$$

Since 39 – 32 is 7, the remainder is 7, as shown.

Another way to use Fingermath to solve $8\overline{)39}$ is to do nearly all the work without pencil and paper. Just count the number of 8s that take you to 32. As before you will count 8 four times. Now to reach 39, you've got to count off the remainder; the numbers from 32 to 39. Keep your press of 32 and begin a new count with "One". Continue counting until you recognize 39 on your fingers.

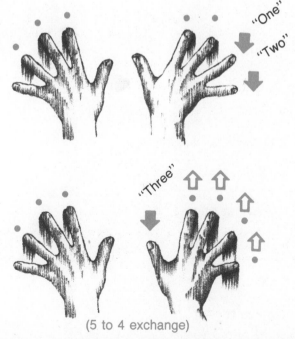

(5 to 4 exchange)

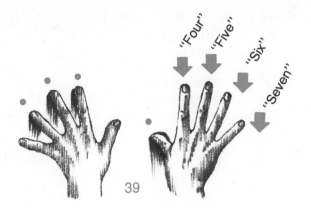

There it is 39!

You have counted 7.

Enter this remainder in your answer.

$$8\overline{)39}^{\,4\ r7}$$

As you will see, you could have taken a more direct route to press from 32 to 39. You had 32 pressed:

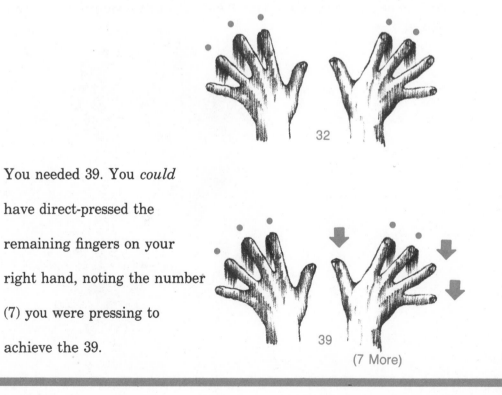

You needed 39. You *could* have direct-pressed the remaining fingers on your right hand, noting the number (7) you were pressing to achieve the 39.

(7 More)

Move along now to the division of a dividend with many digits:

$$6\overline{)2751}$$

You will use the same procedure as before except that your remainders, until the final calculation, will be used as Carryovers.

Watch:

A. You cannot divide 2 by your divisor of 6 because it's smaller than 6. Therefore, combine the first two digits, and try 27.
B. Press 6s up to but not beyond 27, counting how many. You will end with a press of 24 and you will have counted 4. Enter the 4 as shown, above the line.

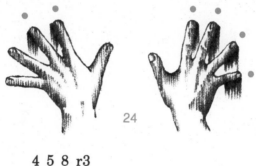

$$\begin{array}{c}4\ \ 5\ \ 8\ \ r3\\ 6\overline{)2\ \ 7\ \ 5\ \ 1}\end{array}$$

C. Keeping 24 pressed, begin a unit-by-unit count until you recognize 27.

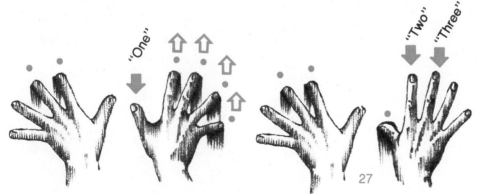

At a press of 27, you will have counted 3.

D. Enter 3 as a carryover into the dividend, in front of the 5, as shown, The 5 is now read as 35 and becomes the next number to be divided by 6.

E. Press 6s up to but not beyond 35, counting how many. You will end with a Press of 30 and you will have counted 5. Enter the 5 as shown following the 4 above the line.

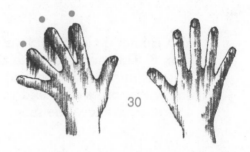

F. Keeping 30 pressed, begin a new count until you recognize 35.

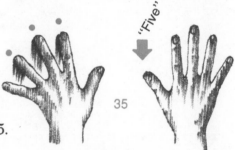

At 35, you will have counted 5.

G. Enter 5 as a carryover into the dividend, in front of the 1, as shown. The 1 now is read as 51 and becomes the last number to be divided by 6.

H. Press 6s up to but not beyond 51, counting how many. You will end with a press of 48 and you will have counted 8. Enter the 8 as shown above the line.

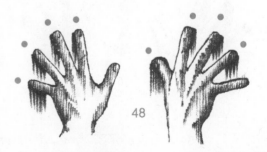

I. Keeping 48 pressed, begin a new count until you recognize 51.

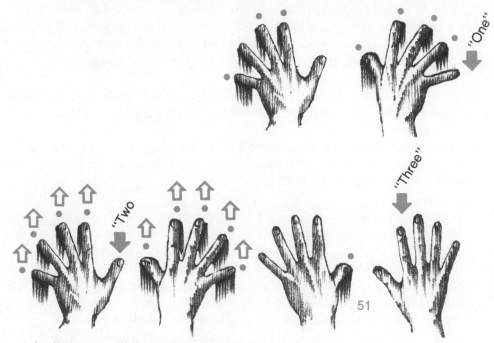

At 51, you will have counted 3.

J. Since there are no more numbers in the dividend, this 3 becomes the remainder. Enter it above the line and mark it with r. You now have your result: 2,751 ÷ 6 = 458 r3. In other words, the number 6 is contained 458 times in 2,751, with a remainder of 3.

The answer in division is called a quotient. A remainder can be 0 or a number greater than 0 (but it must be less than the divisor. Otherwise, you have done your division incorrectly).

Here is the same example you just completed with Fingermath, shown in the form a child learns in school.

$$
\begin{array}{r}
458\ \text{r}3 \\
6\overline{)2751} \\
\underline{24} \\
35 \\
\underline{30} \\
51 \\
\underline{48} \\
3
\end{array}
$$

You could proceed through this familiar form using Fingermath. Just think of the example in pieces. First, solve $6\overline{)27}$, record 4 in the quotient and 24 under the 27. Next subtract 24 from 27 giving you 3. Bring the 5 down from the dividend. Now begin the next piece of division and solve $6\overline{)35}$. Continue this same piece by piece pattern to conclusion.

You can reinforce your division with this usual form by using Fingermath because your multiplication facts always will be handy.

That is the sum and substance of simple division. You keep adding the smaller number (divisor) to determine how many times it is contained in the larger number (dividend).

Following are a couple of pages of simple division examples. Complete each one by yourself, then have someone read each one to you so you get used to hearing a problem and responding with the right finger activities. Listen carefully to the numbers. To make it easier to remember the number you must finally recognize (dividend), as soon as you hear the example, press the dividend and observe it on your fingers: that's the number you'll be aiming to reach. Then clear your fingers and start pressing the divisor, repeatedly, until you recognize the dividend. The number of times you pressed the divisor to reach the dividend becomes your answer (quotient).

Simple Division

This entire page to be repeated orally.

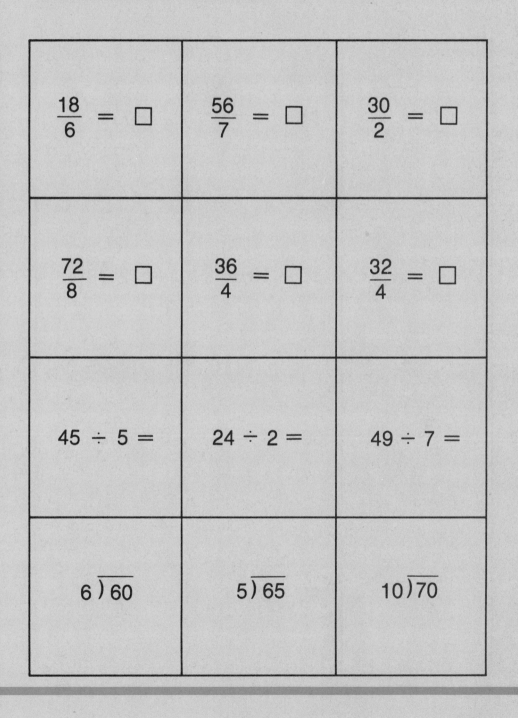

$\frac{18}{6} = \square$	$\frac{56}{7} = \square$	$\frac{30}{2} = \square$
$\frac{72}{8} = \square$	$\frac{36}{4} = \square$	$\frac{32}{4} = \square$
$45 \div 5 =$	$24 \div 2 =$	$49 \div 7 =$
$6\overline{)60}$	$5\overline{)65}$	$10\overline{)70}$

DIVIDE AND CONQUER YOUR FEARS

Simple Division

This entire page to be repeated orally.

$\dfrac{96}{8} =$	$88 \div 8 =$	$7\overline{)35}$
$\dfrac{63}{9} =$	$90 \div 10 =$	$2\overline{)48}$
$\dfrac{78}{6} =$	$65 \div 5 =$	$8\overline{)80}$
$\dfrac{56}{4} =$	$90 \div 6 =$	$4\overline{)76}$

More Addition and Subtraction

Adding Large Numbers: Easier Than It Looks

I'll assume that I'm about the same as everyone else when it comes to making a snap judgment about any activity I face. It is the *number* of challenges I'm likely to meet on a particular venture that seems to govern my appraisal of how simple or how complex it's going to be. When I'm thinking of driving to the local store for a newspaper, I don't even consider it an auto trip. But if I'm contemplating a journey of 200 miles in my car, all kinds of demons come out of the woodwork . . . even though the act of driving and the necessary skills are exactly the same for both trips.

Part of the widespread fear of arithmetic, I believe, comes about in a similar way. Size, or magnitude, seems to become a measure of how simple or difficult an example is. For example, look at

$$\begin{array}{r} 8 \\ +5 \\ \hline \end{array}$$

and there is no great feeling of concern. But the prospect of facing up to something like this:

$$\begin{array}{r} 26,948 \\ 37,585 \\ 48,696 \\ 82,759 \\ 54,362 \\ +99,807 \\ \hline \end{array}$$

can create huge waves of anxiety, particularly in a child. Yet, here again, the procedures and skills are nearly identical to those required to solve the 8 + 5 problem.

I must confess that part of the concern in this situation is justified, because adding up long strings of numbers—even in a single column—has traditionally involved remembering each successive subtotal as you move from number to number. So your attention has to be divided between two constantly shifting objects: it must be fixed momentarily on the total already achieved at any point in the column and it must simultaneously monitor the addition of the next number, which results in a new subtotal to remember.

But you'll now see that, by using Fingermath, this double duty can be reduced to a single one. Your fingers are able to accumulate and simultaneously store each number. You'll never lose your place. The 200-mile drive becomes as simple as a trip around the corner.

You've already discovered, in adding pages of numbers in single columns or strings, that you possess in your fingers a unique mechanism that works for you just like a calculator. Now, by way of another very simple use of that mechanism, you will learn how to handle large blocks of numbers in many columns and rows.

Add up this column:

$$
\begin{array}{r}
2 \\
3 \\
1 \\
2 \\
+1 \\
\hline
\end{array}
$$

Your answer of 9 ends the project!

Now, do this one:

$$
\begin{array}{r}
5 \\
6 \\
3 \\
+7 \\
\hline
\end{array}
$$

Again, when you enter your answer of 21, there is nothing further for you to do.

But suppose this same column is in fact only the units column of a more complicated sum, like this:

$$
\begin{array}{r}
1\,5 \\
2\,6 \\
3\,3 \\
+1\,7 \\
\hline
2\,1
\end{array}
$$

Add it up, and you're faced with a double-digit total even before you reach the second column.

This number 21 contains one unit and two 10s. The two 10s are on your left hand and the one unit is on your right.

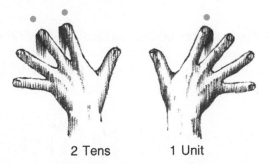

2 Tens   1 Unit

You may enter only one number under any single column, so, under the right units column (which you just added), write the number that is registered on your right hand, the units hand. This number, as you see, is 1. Now look at your left hand. It holds two 10s (20).

These 10s belong with the numbers in the tens column. So write a 2 above that column.

$$
\begin{array}{r}
2 \\
1\,5 \\
2\,6 \\
3\,3 \\
+1\,7 \\
\hline
1
\end{array}
$$

What you've done is carry that 2 over to the tens column. So start calling it a carryover number.

Now, clear your fingers and add up the numbers in the tens column, starting at the top with the carryover number. *Use your right hand.* Even though these numbers are in the tens column, you will add them as you would add units. What you are really adding is that number of tens, but it's the total *number* you want, and it doesn't matter whether the things you are adding are tens or units. Forget about the units column you've already completed. In effect, you have renamed a *new* units column.

Add it up.

$$\begin{array}{r} \boxed{2} \\ 1 \\ 2 \\ 3 \\ + 1 \\ \hline \end{array}$$

This procedure will always be used no matter how many columns must be added in an example. As you begin each new column, add it as if you were adding units, beginning with the carryover number.

$$\begin{array}{r} \boxed{2} \\ 1\ 5 \\ 2\ 6 \\ 3\ 3 \\ + 1\ 7 \\ \hline \end{array}$$

Enter the column total of 9. There is your answer: 91.

Press that whole number:

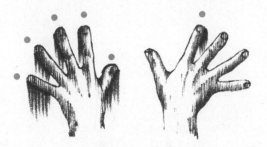

You can see you have nine 10s and one unit.

Try three columns.

```
        T H T U
        H U E N
        O N N I
        U D S T
        S R       S
        A E
        N D
        D S
        S
       ┌─┬─┬─┐
       │3│2│2│
       └─┴─┴─┘
         9 5 6
         2 3 4
         5 4 5
         5 5 5
       + 9 8 5
       ─────────
       3,2 7 5
```

Note that any three-column example works with a units column, a tens column, and a hundreds column. Read *across* and you'll see that your first number, 956, consists of nine 100s, five 10s, and six units. I've also labeled a thousands column, because your final answer is in the thousands.

Add up the units column. Your total should be:

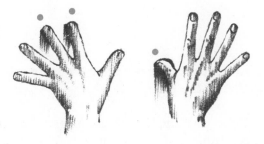

Enter the 5 units under the units column. Register the two 10s as a carryover number, 2, in the box.

Now, clear your fingers and add up the 10s column as though it were units. Your total should be:

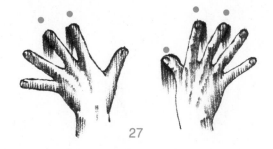
27

```
  3 2 2
  9 5 6
  2 3 4
  5 4 5
  5 5 5
+ 9 8 5
─────────
3, 2 7 5
```

Enter your 7 under the new units column. Carry over 2 (for your two 10s) to the proper box (now at the top of a new 10s column). Just as each column being added must be treated as a new units column, so must each adjacent column to the left be treated as a new 10s column. Keep moving over to the left, repeating the same procedure.

Now, clear your fingers and add up the 100s column as though it were units. Your total should be:

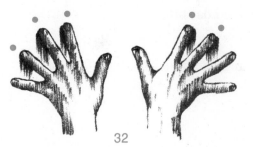
32

Enter your 2 under the new units column. Carry over 3 (for your three 10s) to the box.

Clear your fingers. Never forget to do this before you start adding a new column.

There is only one number in the new units column. Simply enter it at the bottom of that column. Then read your answer: 3,275. (Again, this answer signifies three 1000s, two 100s, seven 10s and five units.)

Do some practice pages now, and then you'll be ready to take a real
shortcut that eliminates the need for boxes and the writing of carryover
numbers altogether.

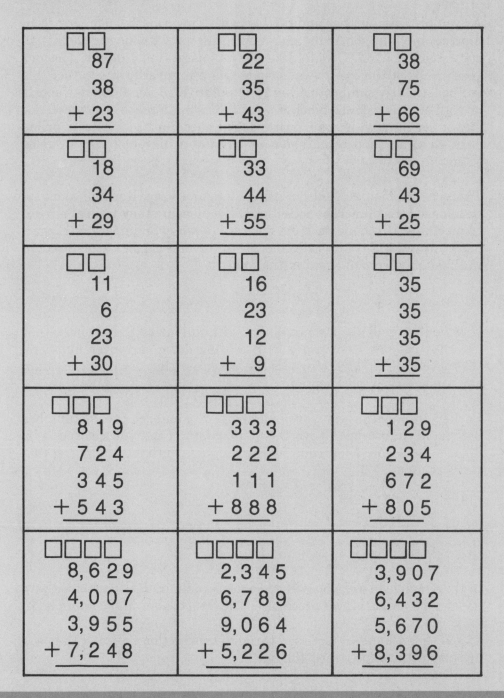

□□ 87 38 + 23	□□ 22 35 + 43	□□ 38 75 + 66
□□ 18 34 + 29	□□ 33 44 + 55	□□ 69 43 + 25
□□ 11 6 23 + 30	□□ 16 23 12 + 9	□□ 35 35 35 + 35
□□□ 8 1 9 7 2 4 3 4 5 + 5 4 3	□□□ 3 3 3 2 2 2 1 1 1 + 8 8 8	□□□ 1 2 9 2 3 4 6 7 2 + 8 0 5
□□□□ 8,6 2 9 4,0 0 7 3,9 5 5 + 7,2 4 8	□□□□ 2,3 4 5 6,7 8 9 9,0 6 4 + 5,2 2 6	□□□□ 3,9 0 7 6,4 3 2 5,6 7 0 + 8,3 9 6

GETTING OUT OF A BOX: Automatic Carryovers

As you progress further and further with Fingermath, each new technique points out shortcuts and simplifications without sacrificing accuracy and control. I would like to emphasize that I do not consider speed to be of the essence, although it is a natural by-product of this method. Clearly similar to other manipulative, motor skills like sports or playing musical instruments, your Fingermath takes on competence and speed through regular practice. But speed in itself is not a prime goal. Accuracy and comprehension are the achievements most to be sought.

There is much to be said for going slow and easy. Rushing along at headlong speed often robs activities of the pleasure they yield at a more comfortable, rhythmic pace. As I've said to many an impatient Fingermath enthusiast, if one wished merely to traverse a dance floor, he could walk quickly from one end to the other.

* * * * *

Now you're ready for a more direct skill, unique to Fingermath, that makes it possible to dispense with writing down carryover numbers. Whenever you're able to bypass the act of writing down information, you are saving wasted motions.

Earlier, in handling carryover numbers, you entered the 10s number (left hand) into a box above the 10s column. Then you cleared your fingers and began a new count of the numbers in that column, treating them as units.

$$
\begin{array}{r}
3\ 9 \\
5\ 6 \\
+\ 7\ 9 \\
\hline
1\ 7\ 4
\end{array}
$$

In the example above, you will add each column in the accustomed way but change the carryover procedure. The total of your first column is 24.

(If you're left handed, just read through this section first and then see my instructions for you at the end.)

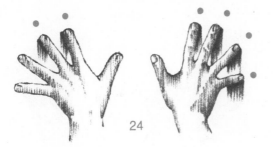

Write down the 4, as always. Your right hand now is clear. Look at your left hand, noting which fingers are pressed. There are two. Press the corresponding fingers on the right hand, at the same time clearing the left.

Your carryover number is now registered on the right hand. You can immediately begin adding the numbers in the next column (being careful not to clear the 2).

The total of this column is 17.

Repeat the procedure as before.

Write down 7 below the column just added, clearing your right hand.

Look at your left hand and observe how many fingers are pressed. There is only 1. Transfer this 1 to the corresponding finger on the right hand (clearing the left).

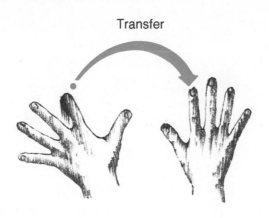

Transfer

Since there are no other numbers in the next column, simply write this 1 into your final answer.

Hereafter, don't write down any carryover numbers. Just transfer them from the left to the right hand.

If you're right handed, skip to the next set of examples on page 283. If you're left handed, read on!

LEFTIES BEWARE

If you're a southpaw, I got ahead of you on that automatic carryover procedure. It's the one Fingermath technique that changes for those who are not right-handed, for the simple reason that you need your left hand to write down the units number, and you can't at the same time hold onto the 10s number. What can you do about it?

Look again at the same example:

$$\begin{array}{r} 3\,9 \\ 5\,6 \\ +\,7\,9 \\ \hline 1\,7\,4 \end{array}$$

You added the first column
and totalled 24.

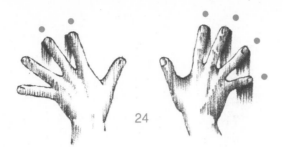

In order to dispense with writing down the carryover number, you must
do two things at once! Try this *very slowly,* as if in slow motion, and
observe your own actions:

You've got 24 pressed. Your pencil is cradled in your left hand. As you
lift up your left hand to write down the 4, transfer its press of 2 fingers
to your right hand, which clears all but those 2 fingers:

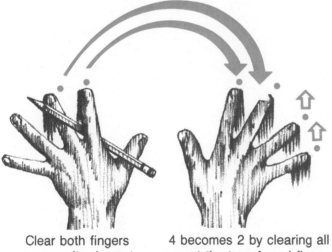

Clear both fingers   4 becomes 2 by clearing all
as you write down 4.   except the transferred fingers.

You are simultaneously transferring the carryover (2) and writing down
the units number (4).

Then continue to add the next column, beginning with the 2 already
registered on your right hand.

Practice this several times using the following double digits first pressing the whole number, then transferring the left-hand fingers to the
right hand and writing down the number that *had* been registered on
the right hand.

16	19
23	32
37	56
42	77
58	68
64	91
71	25
89	55
95	43
13	84
36	18
27	96
59	24
44	83
62	75
81	38
98	52

Go through this entire exercise again. Then tackle the examples that follow, using this identical technique after adding up each column.

Automatic Carryovers

869 457 + 328	246 937 + 582	697 435 + 482
1,689 + 8,534	8,654 + 9,372	3,986 + 4,275
3,841 + 9,065	8,986 + 5,207	5,986 + 3,247
27,897 + 24,342	63,219 + 87,582	46,829 + 37,295
85,275 + 20,745	96,828 + 34,282	14,875 + 89,648
555,555 + 567,895	828,606 + 166,927	398,472 + 842,638

Automatic Carryovers

6,952 8,347 5,906 + 7,219	3,114 8,697 4,208 + 9,696	5,305 6,297 8,454 + 7,396
23,689 47,865 39,072 + 58,345	37,895 64,055 79,621 + 42,050	16,976 23,075 48,959 + 72,524
832,967 246,864 369,541 + 789,623	506,987 321,454 798,632 + 456,789	793,052 906,237 515,748 + 828,367
2,378,496 5,493,862 + 6,589,210	8,639,488 9,210,350 + 7,143,621	6,145,896 3,296,475 + 2,811,032
32,867,489 87,296,340 + 2,468,975	11,370,968 89,547,835 + 62,021,006	96,050,631 267,384 + 4,789,535

MORE ADDITION AND SUBTRACTION

Subtraction: Examples that go beyond two places.

I would commit a serious sin of omission if I ignored the computation of subtraction examples that are a common source of anxiety. These are a familiar looking variety that often seem to defy the laws of borrowing, as I explained them earlier.

You'll recall that I outlined the borrowing or renaming method for examples that involved numbers less than 100. They looked like this:

$$\begin{array}{r} 40 \\ -26 \\ \hline \end{array}$$

Reviewing the renaming procedure quickly, you would, with this kind of example, proceed as follows:

Borrow one of the 4 tens and give it to the 0, making it 10, and renaming the 4 as 3.

Then subtract 6 from 10, giving you 4.

Finally, subtract 2 from 3, giving you 1.
The answer (difference), then, is 14.

I'm sure you now can handle this type of renaming easily. You also can do direct Fingermath subtraction with these examples. However, let's take a look at one that's similar but which makes the renaming procedure more involved and confusing.

$$\begin{array}{r} 8030 \\ -\ 645 \\ \hline \end{array}$$

You'd have no trouble, as shown below, borrowing 1 ten from the 3 and giving it to the 0, making it 10 and renaming the 3 as 2. Then you'd subtract 5 from 10, giving you 5.

$$\begin{array}{r} ^{2\ 10} \\ 80\cancel{3}\cancel{0} \\ -\ 645 \\ \hline 5 \end{array}$$

So far, so good. Then what? How would you make the 2 large enough to be able to accept the subtraction of 4? You can't borrow anything from the 0. In this situation, ordinarily, you must borrow 1 (thousand) from the 8 (making it 7) and give it to the 0 (making it 10). Then you must borrow 1 (hundred) from the 10 (making it 9) and give it to the 2 (making it 12). It would look like this:

$$\begin{array}{r} ^{9\ 12} \\ ^{7\ 10}\ \cancel{2}\ ^{10} \\ \cancel{8}\ \cancel{0}\ \cancel{3}\ 0 \\ -\ 6\ 4\ 5 \\ \hline 7\ 3\ 8\ 5 \end{array}$$

Finally, using Fingermath, you could complete the subtraction of 4 from 12, the subtraction of 6 from 9 and the subtraction of 0 from 7.

As if that weren't complicated enough, things go from bad to worse when an example expands to look like this:

$$\begin{array}{r} 5\,0,0\,0\,4 \\ -\ 2,7\,8\,6 \\ \hline \end{array}$$ minuend
       subtrahend

Those zeros, in particular, make traditional renaming a drudgery. But there is an alternative approach to subtraction known as the Equal Additions method.

Notice in the above example that there is a need to rename every number in the minuend (except the 5) because the corresponding subtrahend numbers are too large to subtract. Before I use the Equal Additions procedure to solve the above problems, let me tell you how this method works.

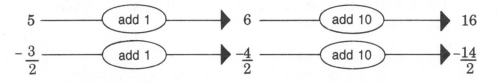

As you can see, if you add the same number to both the minuend and the subtrahend, the difference stays the same. In each of the examples above, the difference is 2.

Now apply this idea to a simple example such as 40 – 26.

$$\begin{array}{cc} \text{tens} & \text{ones} \\ 4 & 0 \\ -2 & 6 \\ \hline \end{array}$$

You can't subtract 6 ones from 0 ones; so add 10 ones to the minuend.

$$\begin{array}{r} {}^{10} \\ 4\cancel{0} \\ -\underline{26} \\ \end{array}$$

Of course, we need to make an equal addition to the subtrahend. However, instead of adding 10 in the ones place, we simply use place value, and add 1 ten in the tens place.

$$\begin{array}{r} \overset{10}{4\cancel{0}} \\ \underset{3}{-\cancel{2}\,6} \\ \hline 1\,4 \end{array}$$

Think:
4 tens + 10 ones

3 tens + 6 ones

Now we subtract: 10 ones – 6 ones is 4 ones.
4 tens – 3 tens is 1 ten.

The difference is 14.

Even though the original numbers have been changed by adding an equal amount both to the minuend and to the subtrahend, the difference remains the same! You can apply the identical procedure to any example. Just use the idea that 10 ones = 1 ten; 10 tens = 1 hundred; 10 hundreds = 1 thousand, etc.

Here's the example I showed you earlier:

$$\begin{array}{r} 50{,}004 \quad \text{minuend} \\ -2{,}786 \quad \text{subtrahend} \\ \hline \end{array}$$

Whenever 10 is added in the ones place of the minuend, 1 is added in the tens place of the subtrahend; whenever 10 is added in the tens place of the minuend, 1 is added in the hundreds place of the subtrahend; and so on.

There is no need to borrow or rename across a whole series of places; simply adjust and subtract only one column at a time.

Step 1

$$\begin{array}{r} 5\,0{,}0\,0\,\overset{14}{\cancel{4}} \\ -2{,}6\,\underset{9}{\cancel{8}}\,6 \\ \hline 8 \end{array}$$

Add 10 ones to minuend

Add 1 ten to subtrahend
Subtract 6 from 14

Step 2

$$
\begin{array}{r}
{\scriptstyle 10\ 14} \\
5\,0,0\,\cancel{0}\,\cancel{4} \\
{\scriptstyle \downarrow} \\
{\scriptstyle 7\ 9} \\
-\ 2,\cancel{0}\,\cancel{8}\,6 \\
\hline
1\ 8
\end{array}
$$

Add 10 tens to minuend

Add 1 hundred to the subtrahend

Subtract 9 from 10

Step 3

$$
\begin{array}{r}
{\scriptstyle 10\ 10\ 14} \\
5\,0,\cancel{0}\,\cancel{0}\,\cancel{4} \\
{\scriptstyle \downarrow} \\
{\scriptstyle 3\ 7\ 9} \\
-\ \cancel{2},\cancel{0}\,\cancel{8}\,6 \\
\hline
3\ 1\ 8
\end{array}
$$

Add 10 hundreds to the minuend

Add 1 thousand to the subtrahend

Subtract 7 from 10

Step 4

$$
\begin{array}{r}
{\scriptstyle 10\ 10\ 10\ 14} \\
5\,\cancel{0},\cancel{0}\,\cancel{0}\,\cancel{4} \\
{\scriptstyle \downarrow} \\
{\scriptstyle 1\ 3\ 7\ 9} \\
-\ \cancel{2},\cancel{0}\,\cancel{8}\,6 \\
\hline
4\ 7,3\ 1\ 8
\end{array}
$$

Add 10 thousands to the minuend

Add 1 ten thousand to the subtrahend

Subtract 3 from 10
Subtract the final 1 from 5

Therefore, since all the minuend numbers are too small (except the 5), your procedure was to

Add 10 to each minuend number except the 5.

Add 1 to each subtrahend number except the first 6.

Then subtract each column with Fingermath.

After practicing awhile there will no longer be a need to write all the little numbers. You can do the work mentally. Here's another example:

Think it through this way:

$$
\begin{array}{r}
5\,0,0\,3\,0 \\
-\ 8,2\,3\,1 \\
\hline
\end{array}
$$

Ones place	$10 - 1 = 9$
Tens place	$13 - 4 = 9$
Hundreds place	$10 - 3 = 7$
Thousands place	$10 - 9 = 1$
Ten Thousands place	$5 - 1 = 4$

Answer: 41,799

Now look at another variety of examples in which all possibilities exist.

Some numbers must be renamed and others must not. Some minuends are larger, some subtrahends are larger and some are equal.

$$\begin{array}{c} \text{G F E D C B A} \\ 1310\,1011 \\ 7,\not{8}\,2\,\not{0},\not{0}\,4\,\not{1} \quad \text{minuend} \end{array}$$

$$\begin{array}{c} 72\,43 \\ \not{6},4\,\not{1}\,\not{8},5\,\not{2}\,8 \\ \underline{} \\ 9\,0\,6,\,5\,1\,3 \quad \text{subtrahend} \end{array}$$

I have labeled the columns from A to G for purposes of explanation.

Column	Minuend	Subtrahend
A	Add 10	Always leave as is
B	Leave as is because the subtrahend is smaller	Add 1
C	Add 10	Leave as is because previous minuend was not renamed
D	Add 10	Add 1
E	Leave as is because the subtrahend is smaller	Add 1
F	Add 10	Leave as is because previous minuend was not renamed
G	Always as is	Add 1

What really is happening here is that adjustments are being made (as in all renaming procedures) so that the minuend is large enough for the subtrahend to be subtracted. If you examine this procedure, you will find that it is very simple.

When you add 10 to the first minuend number, you are adding it in the Units column (A). If you stopped there, the example would be out of balance. (You couldn't add 10 pounds to one side of a scale and have it remain as it was without adding 10 pounds to the other side.) Therefore, when you add 1 to the subtrahend in the adjacent column (B) you are actually adding 1 ten (the same value as 10 units). When you add 10 units to the minuend and 1 ten to the subtrahend you are balancing back to the original position: the net value has not been changed! That explains why you must never adjust a subtrahend in column C, for example, when the minuend in column B was not renamed. Looking at it the opposite way, you must always adjust a subtrahend number when the preceding minuend number *was* renamed. Remember, it only is necessary to think about, adjust and subtract one column at a time. In the problem just outlined, consider A. You must adjust the minuend. Do it and immediately subtract 8 from 11. Then consider B. First balance the subtrahend, noting that it still remains smaller than the minuend. Subtract 3 from 4. Then consider C. Because the minuend in B was not altered, leave C's subtrahend as is. It's larger than C's minuend which must be renamed. Then subtract 5 from 10. Continue to the end, working your way over one column at a time. Practice with the examples I've given you on the following page. You'll find that after working only a few of them, using Fingermath to subtract each column, this kind of example will never bother you again.

Subtraction with Equal Additions

634 −586	703 −649	340 −286	5,030 −4,234	6,002 −4,623
8,393 −6,725	1,036 −527	2,004 −1,869	5,900 −826	63,050 −48,362

MORE ADDITION AND SUBTRACTION

Subtraction with Equal Additions

20,004 −18,126	38,090 −6,247	306,200 −284,763	590,003 −439,215	804,060 −25,127
920,400 −3,428	300,000 −137,204	609,003 −4,785	860,401 −6,284	621,500 −7,000

Advanced Multiplication and Division

Solving Multi-Digit Problems

Something to notice about Fingermath is that the form of an example —that is, the way it looks on paper or the way it sounds—is exactly the way it *always* looked or sounded. There are no special rules to follow in writing down problems. Therefore, you need have no fear that your child will be seeing examples written one way in school and a different way for Fingermath.

The major difference occurs in the way the problem is solved, or the manner in which calculations are performed. Traditional methods assume that various sums and products have been learned by rote. Only to the extent that memory serves you, are you capable of dealing with a varying complexity of problems. With the strength of Fingermath in your hands, however, the absolute need to memorize is eliminated. As I told you earlier, though it is desirable to learn one's facts by heart, the urgency of the need vanishes when one has a personal calculator at hand. Difficulties with memory work, which are at the root of the anxieties about math that many of us carry throughout life, need never trouble anyone who knows Fingermath. For some who have severe learning disabilities, so that they may never be capable of memorizing tables, the ready availability of arithmetic facts through Fingermath is a real boon.

This Fingermath advantage is evident when solving problems of multi-digit multiplication.

Begin with the multiplication of two double-digit numbers, a typically difficult example when one has not memorized multiplication tables.

$$
\begin{array}{r}
\boxed{2} \\
\boxed{4} \\
5\,7 \\
\times 3\,6 \\
\hline
3\,4\,2 \\
1\,7\,1 \\
\hline
2,0\,5\,2
\end{array}
$$

Follow me through this example, step by step. Note that I am using the traditional method of solving the problem, with the sole difference that I am depending on Fingermath instead of using memorized figures.

A. Multiply 6 × 7
(Add 7 six times)

Result =

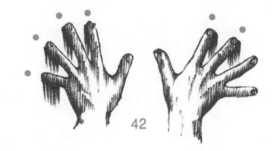

Read your fingers. (Left hand has 40, the Right has 2.)

B. Write 2 below the line under the units column.
Write 4 above the 10s column as a carryover.
Clear your fingers.

C. Multiply 6 × 5 (Add 5 six times)

Result =

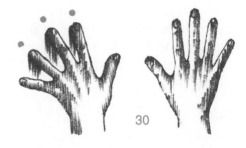

ADVANCED MULTIPLICATION AND DIVISION

D. Hold the 30. Add the carryover of 4.

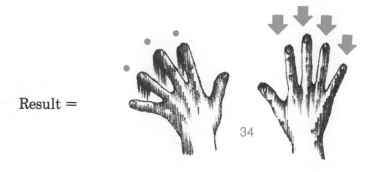

Result =

34

E. Enter 34 below the line, next to the 2. Clear your fingers. That completes the multiplication of 6 × 57. (Result = 342, as entered below the line) Now continue by multiplying the next set, 3 × 57, as follows:

F. Multiply 3 × 7 (Add 7 three times)

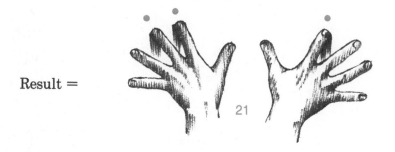

Result =

21

(Left hand has 20, the right hand has 1.)

G. Write 1, as shown, under the 4 in 342 (Actually, you are lining it up with the 3 in 36, which is the number you are presently multiplying.) Write 2 above the 10s column as a carryover. (Cross out the previously carried 4 so you don't get mixed up.) Clear your fingers.

H. Multiply 3 X 5 (Add 5 three times)

Result =

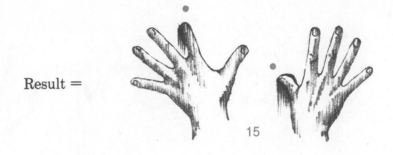

I. Hold the 15. Add the carryover of 2.

Result =

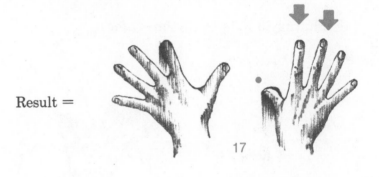

J. Enter 17 next to the 1.
That completes the multiplication of 3 X 57.

(Result = 171, entered as shown.)

K. Add the partial products 342 and 171 as you have learned to do with Fingermath. Your final answer is 2,052.

* * * * *

Now let me show you why it's a distinct advantage when you finally do memorize your multiplication facts for all multipliers from 1 to 9.

It then becomes possible with Fingermath to dispense with writing down carryovers. You have already learned this technique for addition. Now put it to use in Multiplication.

$$
\begin{array}{r}
5\ 7 \\
\times\ 3\ 6 \\
\hline
3\ 4\ 2 \\
1\ 7\ 1 \\
\hline
\end{array}
$$

A. Knowing 6 × 7 = 42, PRESS 42

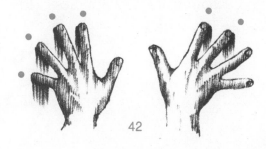

42

B. Enter 2 below the units column and transfer the carryover of 4 from your left hand to your right.

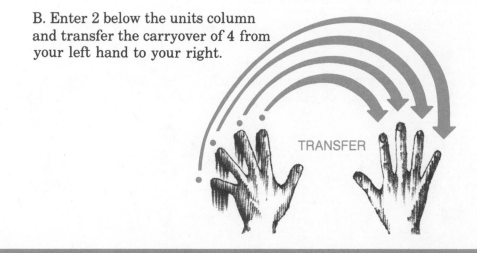

TRANSFER

C. Knowing 6 × 5 = 30, add this 30 to the 4 already established on your right hand.

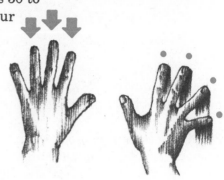

```
    57
  × 36
  ─────
   342
   171
```

D. Enter the result, 34, below the line next to the 2. Clear your fingers.

E. Repeat the process with the multiplier 3.
Knowing 3 × 7 = 21, PRESS 21.

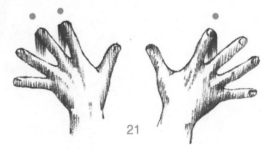

21

F. Enter 1 (below the 4) and transfer the carryover of 2 from your left hand to your right.

TRANSFER

300

G. Knowing $3 \times 5 = 15$, add this 15 to the 2 already established on your right hand.

Enter the result, 17, next to the 1.

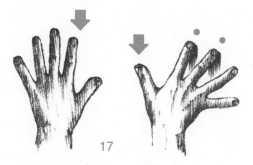

17

H. Add up partial products, as before, for the final answer.

Try each of the examples that follow. You should be able to handle these whether your facts are memorized or not. If you're coming up with incorrect answers, go back to the beginning of this section (page 295) and review all the procedures. Do not move ahead until you are sure you understand multidigit multiplication.

Double Digit Multiplication

29 X12	38 X25	46 X59
87 X38	79 X 4	62 X46
59 X95	93 X27	24 X11
38 X68	48 X22	66 X29
71 X38	28 X40	56 X56

Multidigit Multiplication

367 X14	522 X11	684 X37
829 X25	621 X58	729 X33
406 X23	915 X37	628 X41
1,692 X24	3,847 X 8	2,692 X14
3,833 X45	2,468 X32	6,969 X59

SOLVING MULT-DIGIT PROBLEMS

The next subject, logically, would be long division. But to be able to handle this normally troublesome fellow with simple Fingermath procedures, you must know how to calculate beyond 100.

At last! This is what you've been waiting for. I can't delay it any longer, so here it is.

Outer Limits
Breaking the 100s Barrier

There is a widespread misconception about the limited capacity of the hands in Fingermath. It is assumed that because the number 99 uses up all fingers, you cannot go farther. The fact is that there is no limit to the numbers that can be accumulated. All you have to learn is how it's done. You use your head, literally.

You've probably seen some of the children I've had on TV: six- or seven-year-olds who finger their way through long strings of numbers being called out, and who finish with totals in the hundreds. These kids use their heads. Here's how:

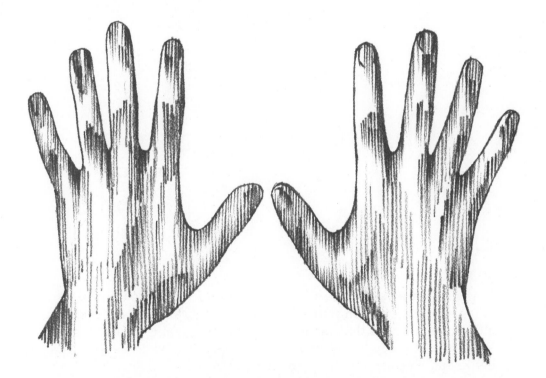

The All-Clear Position Represents 100, 200, 300 etc.

First of all, they follow the same procedures of manipulation that you've already learned. That is, they press numbers as usual, observing all the rules of sequences and exchanges. The only change is in the experience of breaking the 100s barrier.

The number 100 itself is represented in the same way as 0 (zero): all fingers are cleared, in a ready position.

Going from 99 to 100, then, works like this:

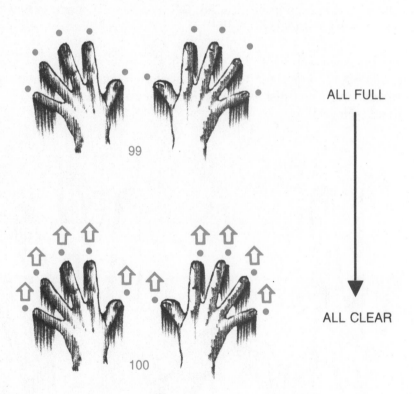

99

100

ALL FULL

ALL CLEAR

In the step from all full to all clear, one unit is added to 99. You've got to register that event in your mind. At the same time, you have acquired a fresh set of hands that can go on to accumulate and store another 99. When you're ready to deliver your answer, refer to your mental storage bank, and be sure to include the first 100.

So far, so good. You can go from 99 to 100 in a one-unit step and then move on. But what if you're adding (or multiplying) numbers larger than 1? Simple.

Since you are proficient with 9s, try to follow me in jumping from a press of 99 past the 100s barrier in one step. I already mentioned that you'll employ familiar sequences and exchanges.

PRESS 99

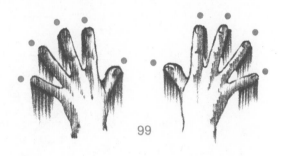

PRESS 9 more

Use the 10 – 1 method (see page 00),
On the left hand:
To go from 90 to 100 (a press of 10)
Clear the entire hand
Result: 100 KEEP IT IN YOUR HEAD
On the right hand:
Clear 1

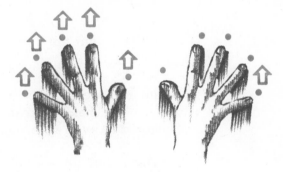

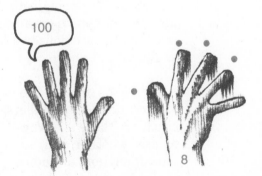

Result (Read your mind *and* your fingers):
108

Now continue:

PRESS 9 MORE

(Use the 10–1 press)
Keep the 100 in your head

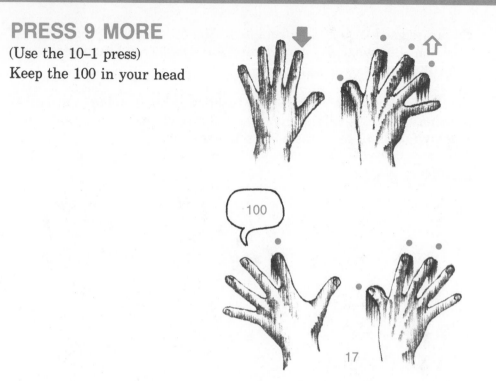

Result (Read your mind *and* your fingers):
117

Continue to accumulate 9s, or any other numbers, to the next 100s barrier. Pass that and keep moving, remembering that you now have 200 stored away in your memory bank.

And that clears up one mystery.

Before going on to the examples on the following two pages, do an entire sequence of each single-digit number from 2 to 9, using the same manipulations that you have already mastered for multiplication. But now, on reaching the previous two-hand limit, go beyond it, passing the 99 mark as illustrated above.

Always keep in mind the various methods available for an instant press of each of these numbers. If one method doesn't work in a specific situation, there is always one that will work. Knowing instantly which to use, whether you are repeating the same number over and over (multiplication) or adding a variety of numbers, is a matter of practice. If you have reached the point where you automatically perform a repeated series of a single number below 99 but you have difficulty recognizing the different press-and-clear combinations above 100, simply go back and review them. I guarantee they will come back to you quickly.

A key point to keep in mind is that crossing the 100s barrier—that is, moving from a fully pressed left hand combined with a right hand that does not have room for the next number—always involves clearing the left hand—immediately adding a value of 10.

When you have finished the entire series of single digits, carrying each at least to 200, try your hands at the following examples.

$95 + 6 = \square$

PRESS 95

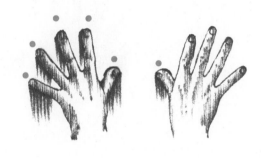

PRESS 6 MORE

You only have 4 available on your right hand. You *must* choose the 5 + 1 method of pressing 6. Therefore, simultaneously press 10 and clear 5 (exchange) and add 1 more.

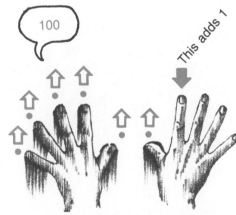

Result: 101   This adds 10   This clears 5

CLEAR YOUR FINGERS.

Here are some more:

$$79 + 8 + 8 + 8 = \square$$

On this example, after the second 8 you are pressing 95. For the last 8 you still must press a 5 exchange + 3.

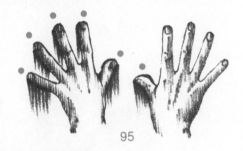

95

For your 5 exchange, add 10 by clearing your left hand and clear 5 by clearing your right thumb. Then simply add 3 more.

Now, do all the examples on the next page.

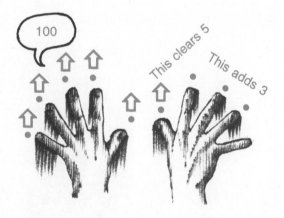

Passing 99 - Exercises

This entire page to be repeated orally.

$82 + 10 + 5 + 8 = \square$ $94 + 7 + 7 = \square$	$79 + 9 + 10 + 8 = \square$ $96 + 10 + 7 = \square$
$96 + 6 + 6 + 6 = \square$ $87 + 9 + 4 = \square$	$97 + 7 + 7 = \square$ $95 + 5 + 6 = \square$
$83 + 7 + 7 + 7 = \square$ $91 + 9 + 9 = \square$	$89 + 9 + 9 = \square$ $76 + 20 + 10 = \square$
$88 + 3 + 3 + 6 = \square$ $89 + 5 + 5 + 5 = \square$	$82 + 20 + 80 + 16 + 9 = \square$ $93 + 9 + 90 + 9 = \square$
$93 + 6 + 6 = \square$ $84 + 10 + 10 = \square$	$87 + 8 + 5 + 95 + 5 = \square$ $78 + 22 + 96 + 4 = \square$

Long Division

I'm not embarrassed to admit that the usefulness of Fingermath is greatest in solving problems that require *basic* computations. You will reach a point in arithmetic where your finger skills cannot serve you, simply because the two-hand working limit is 99. This impedes calculations involving numbers of three or more digits.

Therefore, you can do long division provided the divisor does not exceed two digits.

There is one other if. You can do long division if you practice enough so that you can press double-digit numbers in sequence. And this is not a terribly sophisticated skill. Seven- and eight-year-olds are doing it with speed and accuracy. Naturally, you will find yourself breaking through 100s barriers at a rapid pace.

Here is an example. Add (multiply) 21s, six times.

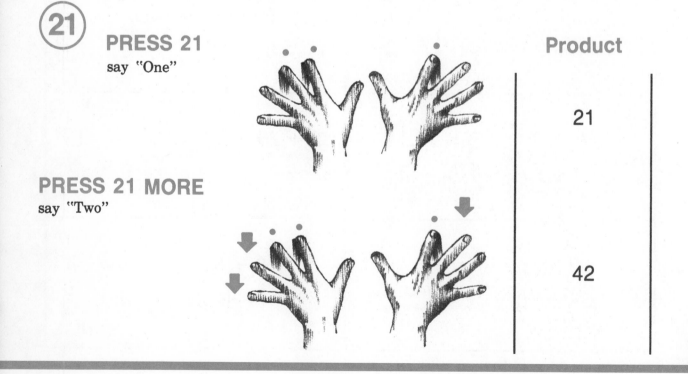

(21)

PRESS 21
say "One"

Product

21

PRESS 21 MORE
say "Two"

42

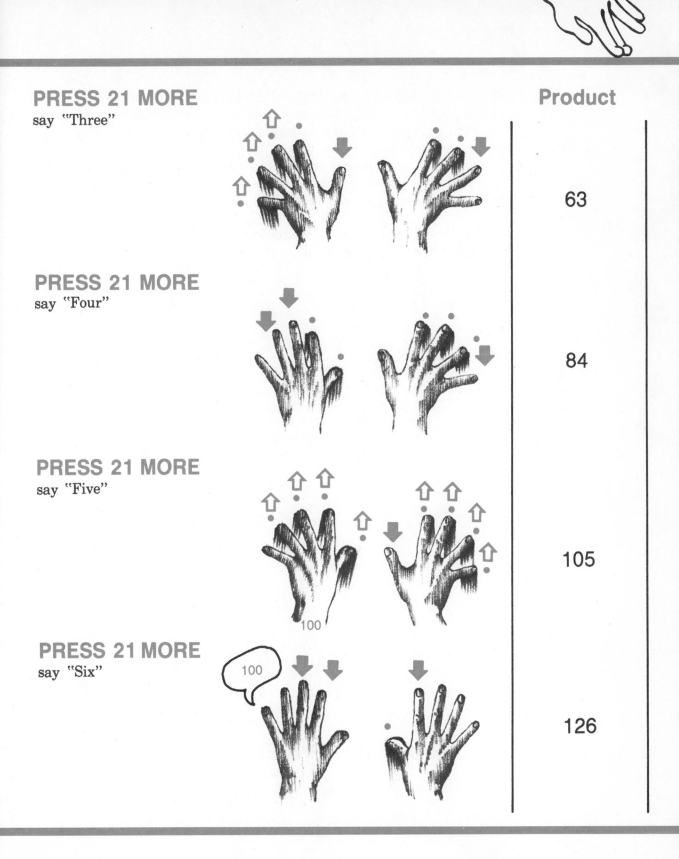

PRESS 21 MORE
say "Three"

PRESS 21 MORE
say "Four"

PRESS 21 MORE
say "Five"

PRESS 21 MORE
say "Six"

Product

63

84

105

126

100

You can do this with any double-digit number. And once you become accomplished (even at a very slow pace) you will be prepared to do long division, but in a much shorter way than you've ever tried or ever seen.

You are going to employ the same procedures that you used for calculating and entering numbers in simple division. Instead of a one-digit divisor, however, you will have two digits.

Since I just finished illustrating the pressing pattern for 21s, that's an excellent divisor to start with.

$$\begin{array}{r} 1\ \ 2\ \ \ 8\ r\ 5 \\ 21\overline{)2\ \ 6\ {}_5 9\ {}_{17}3} \end{array}$$

A. **PRESS 21**

Say "One"

Your next press would take you beyond the 26 you need, so write down 1 above the line. Re-pressing 21, begin a unit-by-unit count until you recognize 26.

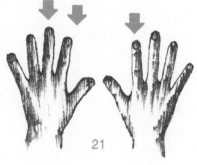

Just 5 does it!

Enter this 5 as a carryover in front of the 9 in the dividend, as shown; it becomes 59.

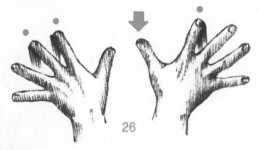

CLEAR YOUR FINGERS.

B. PRESS 21s

Count again from "One," and press 21s up to but not beyond 59. You will reach a press of 42 on a count of "Two." Enter 2 above the line, as shown.

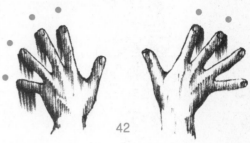

42

Re-pressing 42, begin a units count until you recognize 59. 17 does it. Enter 17 as a carryover, in front of the 3, as shown; it becomes 173.

17 MORE

CLEAR YOUR FINGERS.

C. PRESS 21s

Count from "One" and press 21s, up to but not beyond 173. You will reach a press of 168 on a count of "Eight." Enter 8 above the line, as shown.

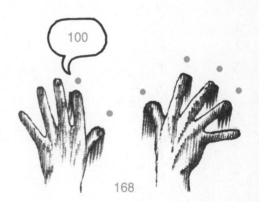

100

168

Repressing 168 (68) count until you recognize 173 (73).

```
        1 2 8 r 5
21) 2 6 9 3
        5 17
```

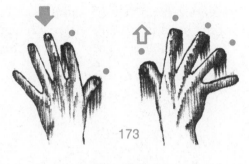

173

(5 exchange)

5 does it!

Enter 5 above the line as a remainder (r).

There is no shorter route to long division. But I must repeat that you can use this method only with a single-digit or a double-digit divisor. Even so, isn't it great to know you, or your child, can go this far toward conquering the fears you've had about division?

ADVANCED MULTIPLICATION AND DIVISION

Long Division

11)2,323	12)3,849	13)4,692
21)2,568	22)3,986	25)6,055
15)3,330	16)1,526	12)3,648
20)69,350	11)21,889	10)69,420
24)53,962	12)64,968	31)63,285
11)39,689	15)36,849	22)33,212
13)9,654,374	12)3,849,678	20)1,145,990

Combining Fingermath Skills

$26 - 11 + 43 - 6 = \square$	$8 \times 9 + 5 - 25 = \square$	$88 \div 8 \times 6 + 60 = \square$ *NOTE:* Count how many 8s are contained in 88(11). Then begin a new calculation of 11 × 6. Finally, add 60.
$56 \div 7 \times 14 + 52 = \square$	$108 \div 6 \times 3 + 50 = \square$	$134 + 27 - 42 = \square$

Division: The Long & The Short of It

$23\overline{)39,645}$	$11\overline{)24,689}$	$9\overline{)148,632}$
$\dfrac{147}{5} = \square$	$136 \div 8 = \square$	$4\overline{)169,723}$
$\dfrac{217}{9} = \square$	$87 \div 3 = \square$	$115 \div 7 = \square$
$16\overline{)348,696}$	$15\overline{)114,835}$	$\dfrac{130}{6} = \square$

Mixed Calculations

$$
\begin{array}{r}
-2,385 \\
3,967 \\
-4,508 \\
5,823 \\
\hline
\end{array}
$$

Here's the kind of arithmetic you can really have fun with. You have probably seen the kids I've had on TV zipping through examples like this. And you undoubtedly wondered how they were able to combine plus and minus entries both in one example in one continuous series of manipulations.

The secret lies in knowing how to set up the problem in advance. From that point, it's simple Fingermath all the way.

Customarily, to solve the above problem, you would separately add up the two minus rows and subtract that sum from the sum of the plus rows, like this:

MINUS	PLUS	DIFFERENCE	
2,385	3,967	9,790	
+ 4,508	+ 5,823	− 6,893	
6,893	9,790	2,897	ANSWER

That's the long way around.

I'll show you the easy way. It requires a mental adjustment which, when you're first learning—you will write down. Look at the first (units) column. If you tried to add and subtract each number in order as it appears on a minus or plus row, you would always start out, and frequently end up, with a minus (negative) amount. So, you're going to compensate for that problem by *over*-adding to each column. The amount you will add is determined by the number of minus rows there are in the problem. In order not to throw off the true value of the example, however, you'll take away the same amount. Let me show you what I mean.

$$
\begin{array}{r}
(-2)(20) \\
-2,385 \\
3,967 \\
-4,508 \\
\underline{5,823}
\end{array}
$$

In this example, there are two minus rows, so I have used the formula that follows. I have added 20 to the first (units) column. To balance that over-addition of 20, I have subtracted two 10s (–2) from the 10s column. The net result is zero. If you add 20 pennies and subtract 2 dimes, you have really changed nothing. Yet you *have* made it possible to calculate. (I'll explain later how to compensate for more than two minus rows.)

Now you'll recall that I've insisted throughout this book that Fingermath methods are consistent. When you do something once, you must always do it. Therefore, you are obliged to continue these 20 – 2 entries for *all* columns.

When you finish, your example looks like this:

$$
\begin{array}{r}
(-20)\ (20)\ (20)\ (20) \\
(-2)\ (-2)\ (-2)\ (-2) \\
-2,\ 3\ \ 8\ \ 5 \\
3,\ 9\ \ 6\ \ 7 \\
-4,\ 5\ \ 0\ \ 8 \\
5,\ 8\ \ 2\ \ 3
\end{array}
$$

For every 20 you have added in a column, you have compensated by subtracting 2 in the next column to the left. Once again, the net result of these entries is zero.

Now Fingermath goes to work.

Calculate down each column, carefully observing whether the number occupies a plus row or a minus row. (The minus signs to the left of the

example identify minus rows. The absence of a sign means a plus row or number.)

When a number is plus, add it; when a number is minus, subtract it. Carryovers are entered or transferred automatically on the fingers, as always.

Here goes:

First column

PRESS 20

Subtract 5

Count back by units

Say "One"

"one"

(20 to 19)

"five" "four" "three" "two"

"Two, Three, Four, Five"

Add 7

(EXCHANGE + 2 . . . REMEMBER?)

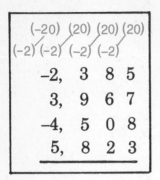

(−20)	(20)	(20)	(20)
(−2)	(−2)	(−2)	(−2)
−2,	3	8	5
3,	9	6	7
−4,	5	0	8
5,	8	2	3

Subtract 8

Count back by units

Say "One, Two"

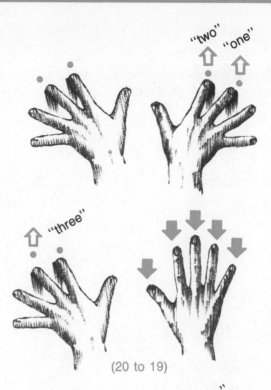

"Three"

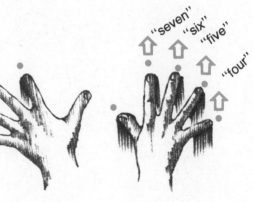

(20 to 19)

"Four, Five, Six, Seven"

"Eight"

(5 to 4)

322

Add 3

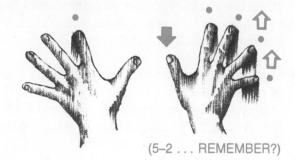

(5–2 . . . REMEMBER?)

Read your total: 17

Enter 7 below the line. Transfer the carryover of 1 to your right hand and proceed in the same manner to find the sum of the other columns. Be certain to observe carefully whether each row is minus or plus; that's all you have to remember.

When you become more experienced, you will be able to skip writing down 20s and –2s across the top of the columns. It's very easy to do this operation mentally—you always merely

> Add 20 to the Units Column
> Subtract 2 from the farthest column to the left (which contains no numbers)
> Add 18 (20 – 2) to all columns in between.

There is one other important fact that will make it possible to calculate even more rows of numbers in a mixed calculation.

The figures 20 and –2 are determined by the number of minus rows in the example. In the above exercise you had two such rows. Say you had three minus entries, like this:

$$
\begin{array}{r}
-\ 2\ 3\ 9 \\
-\ 4\ 8\ 6 \\
8\ 3\ 5 \\
-\ 1\ 6\ 7 \\
\hline
3\ 9\ 4
\end{array}
$$

Then you use 30 and –3 as your compensatory numbers; for four minus rows, use 40 and –4, etc.

It makes no difference in what sequence the plus or minus rows appear or relatively how many there are of each. Just be sure to observe their position carefully when performing your calculation.

One more point. When you're setting up an example like this, the sum of the minus rows must not be more than the sum of the plus rows. You can check this very quickly. Just add up the minus numbers in the extreme left column; this total must be less than the sum of the plus numbers in that column.

Now go back over the example I've explained in detail to be sure you understand the procedure.

Then try the examples on the next pages.

<table>
<tr><td>

(20) (20) (20)
(–2) (–2) (–2)
```
    5  6  7
 –  2  3  4
    3  8  9
 –  1  9  5
```

</td><td>

(30) (30) (30)
(–3) (–3) (–3)
```
    8  9  6
    1  4  7
 –  2  9  5
 –  3  0  2
 –  5  1  6
```

</td></tr>
<tr><td>

( ) ( ) ( )
( ) ( ) ( ) ( )
```
   8, 9 2 3
 – 3, 4 6 7
   5, 8 9 5
 – 4, 6 5 8
```

</td><td>

```
 –  4 6 9
    8 2 2
    5 0 6
 –  2 3 7
 –  1 0 8
```

</td></tr>
<tr><td>

```
   4, 3 2 7
 – 1, 8 9 5
 – 2, 6 5 9
   5, 8 7 8
```

</td><td>

```
   3, 8 0 6
 –    9 2 4
 – 1, 0 7 9
      5 8 8
```

</td></tr>
<tr><td>

```
 – 3, 5 8 5
   9, 6 2 7
   5, 3 8 5
 – 2, 9 6 4
```

</td><td>

```
 – 5, 2 9 6
   8, 0 0 7
      1 4 8
   5, 6 9 2
```

</td></tr>
<tr><td>

```
 – 2 3, 4 6 8
   8 9, 0 0 6
 – 1 5, 9 5 0
   3 8, 2 1 5
```

</td><td>

```
   6 4, 9 0 7
 –  9, 3 2 0
        4 9 6
     5, 0 1 5
 – 1 2, 5 8 9
```

</td></tr>
</table>

Missing Addends

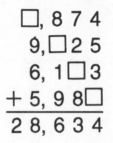

$$\begin{array}{r} \square,\,8\,7\,4 \\ 9,\square\,2\,5 \\ 6,\,1\,\square\,3 \\ +\ 5,\,9\,8\,\square \\ \hline 2\,8,\,6\,3\,4 \end{array}$$

At first glance, this is a real puzzle. It is the opposite of what you normally find in an addition problem. You know the answer but you have to figure out the question! In other words, you have to find what numbers to put in the empty boxes in order to make the answer correct.

With Fingermath, a child of seven can find the missing numbers. Just follow this simple procedure:

A. Add the numbers above the line in the first (units) column. The total is 12. Hold onto it.

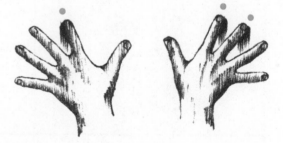

B. Note that the number in the answer below the first column is 4. Ask yourself what number you must count up to, from the 12 already pressed, in order to reach a number ending with 4. That number is 14. Begin a new count, starting with "One," until you recognize 14.

Your count is "two."

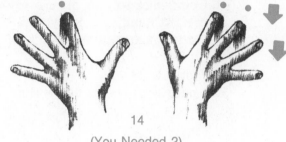

14

(You Needed 2)

C. Enter this 2 in the units box ☐.

D. Carryover the left-hand press to the corresponding fingers of the right hand (as you learned to do much earlier).

E. Add up the next column. (Don't forget to start with your carryover of 1.)

The total is 18. Hold onto it.

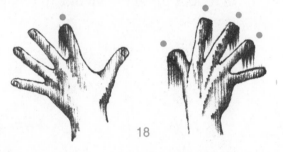

18

F. Look at the 3 below that column.

What number can you count up to from 18 that will end with 3? Right: the number is 23. Begin a new count starting with "One," until you recognize 23. (If you're really sharp, you'll see that a 5 exchange takes you right there!)

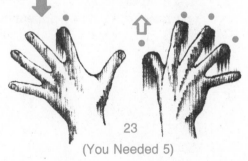

23

(You Needed 5)

```
 6,8 7 4
 9,6 2 5
 6,1 5 3
+5,9 8 2
 2 8,6 3 4
```

G. Enter 5 in the box ☐ .

H. Going back to your press of 23, transfer your left-hand press to your right hand as a carryover.

I. Add the next column. (Don't forget to start with your carryover of 2.)

The total is 20.

You need 26, because below that column a 6 appears.

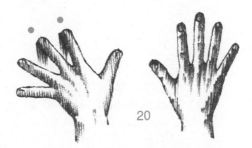

J. Count up to the missing 6 and enter it in the box ☐

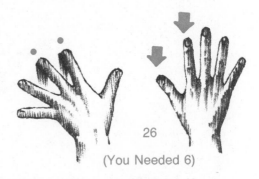

K. Carryover the left-hand press of 2 to the right hand.

L. Add up the last column. (Don't forget to start with your carryover of 2.)

M. The total is 22.

You need 28 (already appearing below the line).

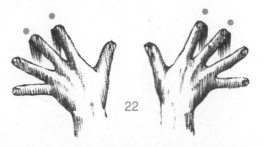

Step 2

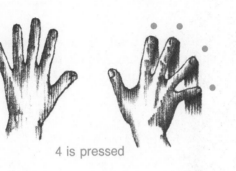

You are required to multiply twice, in the form of an X.

First, keeping 4 (carryover) pressed,

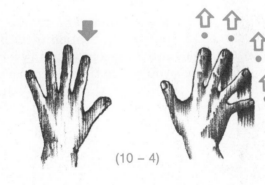

4 is pressed

Multiply 6 × 4.

(Add 4 six times *or* do what requires fewer manipulations, as shown, add 6 four times.*)

Say "One"

(10 − 4)

*This accumulation of 6s on top of the already established 4 is simple if you practiced the exercises outlined on page 225.

N. Count up to 28.
The missing number is 6.
Enter 6 in the box ☐

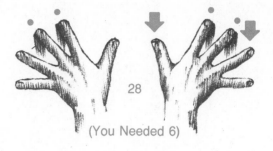

28

(You Needed 6)

Go through this example once more so that you fully understand the procedure. Then try some new ones.

☐,863 9,☐25 7,4☐7 +5,63☐ ———— 26,493	9☐,639 ☐5,420 67,8☐1 89,☐48 +20,63☐ ———— 348,034
5,☐69 ☐,784 9,62☐ +8,5☐1 ———— 29,829	82,97☐ 6☐,045 16,7☐9 53,☐92 +☐5,780 ———— 300,479

3☐9,654 ☐84,210 69☐,587 100,☐32 567,8☐9 + 825,96☐ 3,111,111	896,38☐ 429,2☐5 955,☐62 30☐,798 8☐9,503 +☐10,609 4,345,678
609,34☐ 817,9☐6 742,☐58 98☐,567 2☐7,890 +☐38,805 3,555,555	3☐4,679 ☐98,742 6☐,590 38,6☐7 924,☐95 +862,04☐ 3,255,783
86,95☐ ☐3,842 7☐,514 + 29,☐86 279,1☐8	506,☐9☐ 3☐5,867 90,425 +☐62,739 1,25☐,7☐8
4☐,069 ☐2,587 84,☐88 69,2☐0 + 71,69☐ 356,789	397,54☐ 286,☐93 52☐,304 9☐1,867 ☐56,920 + 604,3☐9 3,548,942

One-Liners
(No Joke)

Well, now I really have a surprise for you, and it's made possible by the capacity of Fingermath to accumulate numbers. I'm talking about multidigit multiplication in One Line.

$$\begin{array}{r} 47 \\ \times 36 \\ \hline 1,692 \end{array}$$

That's a typical example. And that's how it will look when you finish calculating it. No Carryovers. No Partial Products to add. Nothing except your one-line answer. All the time-consuming steps can be eliminated because, with Fingermath, your fingers will store both carryovers and partial products. Talk about amazing your friends—here's a skill that will!

It's all very scientific and very simple. There's only one catch. You should know your multiplication facts if you want to completely dispense with all entries except your answer. With repeated, daily Fingermath practice, you will in time learn these facts by heart. So be patient. Meanwhile, you can run through the following step-by-step procedure, which works with every example of double-digit multiplication, without exception.

Step 1

Multiply 6 X 7. The result is 42.

Enter 2 below the line, clearing the right hand.

Transfer the four pressed fingers on the left hand to the corresponding fingers on the right hand (carryover).

$$\begin{array}{r} (42) \\ 47 \\ \times 36 \\ \hline 2 \end{array}$$

"Two"

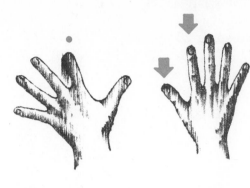

"Three"

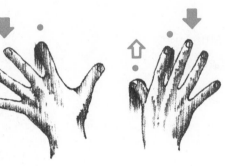

"Four"

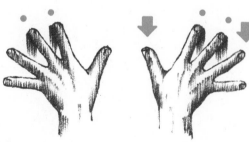

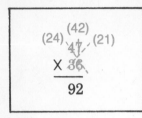

The result is 28.

Keeping 28 pressed,

28 is pressed

Multiply 3 × 7
(add 7 three times)

Say "One"

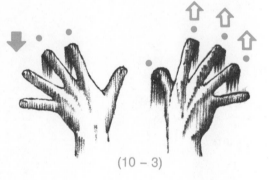

(10 − 3)

"Two"

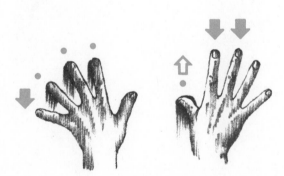

(exchange + 2)

"Three"

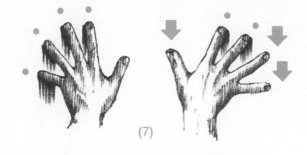

(7)

You have completed multiplying both parts of Step 2 and the result is 49. Enter 9 below the line, clearing the right hand. Transfer the four pressed fingers on the left hand to the corresponding fingers on the right hand (carryover).

Step 3

Keeping 4 pressed,

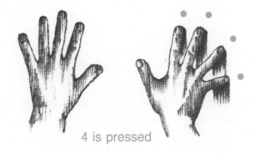

4 is pressed

Multiply 3 X 4
(Add 4 three times)

Say "One"

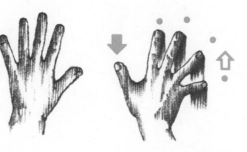

$$
\begin{array}{r}
(12)\ (42) \\
(24)\ \cancel{47}\ (21) \\
\times\ \cancel{36} \\
\hline
1{,}692
\end{array}
$$

"Two"

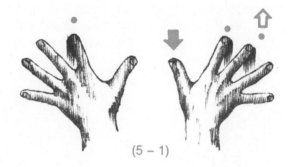

(10 − 6)

"Three"

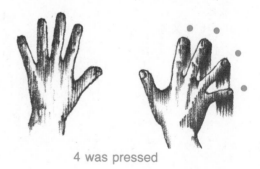

(5 − 1)

The result is 16. This was your last step, so write down 16 below the line. The final product is 1,692.

This identical pattern can be applied to the multiplication of any pair of double-digit numbers. Naturally, once you know your multiplication tables by heart you can really zip through this procedure. For example, look back at Step 2.

4 was pressed

336

N. Count up to 28.
The missing number is 6.
Enter 6 in the box ☐

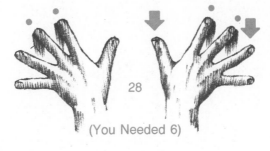

28

(You Needed 6)

Go through this example once more so that you fully understand the procedure. Then try some new ones.

☐,8 6 3 9,☐2 5 7,4☐7 + 5,6 3☐ —————— 2 6,4 9 3	9☐,6 3 9 ☐5,4 2 0 6 7,8☐1 8 9,☐4 8 + 2 0,6 3☐ —————— 3 4 8,0 3 4
5,☐6 9 ☐,7 8 4 9,6 2☐ + 8,5☐1 —————— 2 9,8 2 9	8 2,9 7☐ 6☐,0 4 5 1 6,7☐9 5 3,☐9 2 + ☐5,7 8 0 —————— 3 0 0,4 7 9

3□9,654 □84,210 69□,587 100,□32 567,8□9 +825,96□ 3,111,111	896,38□ 429,2□5 955,□62 30□,798 8□9,503 +□10,609 4,345,678
609,34□ 817,9□6 742,□58 98□,567 2□7,890 +□38,805 3,555,555	3□4,679 □98,742 6□,590 38,6□7 924,□95 +862,04□ 3,255,783
86,95□ □3,842 7□,514 +29,□86 279,1□8	506,□9□ 3□5,867 90,425 +□62,739 1,25□,7□8
4□,069 □2,587 84,□88 69,2□0 + 71,69□ 356,789	397,54□ 286,□93 52□,304 9□1,867 □56,920 +604,3□9 3,548,942

One-Liners
(No Joke)

Well, now I really have a surprise for you, and it's made possible by the capacity of Fingermath to accumulate numbers. I'm talking about multidigit multiplication in One Line.

$$\begin{array}{r} 4\,7 \\ \times\,3\,6 \\ \hline 1,6\,9\,2 \end{array}$$

That's a typical example. And that's how it will look when you finish calculating it. No Carryovers. No Partial Products to add. Nothing except your one-line answer. All the time-consuming steps can be eliminated because, with Fingermath, your fingers will store both carryovers and partial products. Talk about amazing your friends—here's a skill that will!

It's all very scientific and very simple. There's only one catch. You should know your multiplication facts if you want to completely dispense with all entries except your answer. With repeated, daily Fingermath practice, you will in time learn these facts by heart. So be patient. Meanwhile, you can run through the following step-by-step procedure, which works with every example of double-digit multiplication, without exception.

Step 1

Multiply 6 X 7. The result is 42.

Enter 2 below the line, clearing the right hand.
Transfer the four pressed fingers on the left hand to the corresponding fingers on the right hand (carryover).

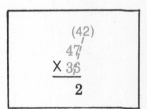

Step 2

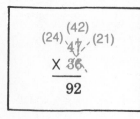

You are required to multiply twice, in the form of an X.

First, keeping 4 (carryover) pressed,

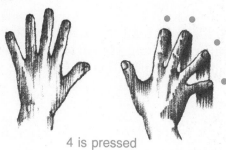

4 is pressed

Multiply 6 × 4.

(Add 4 six times *or* do what requires fewer manipulations, as shown, add 6 four times.*)

Say "One"

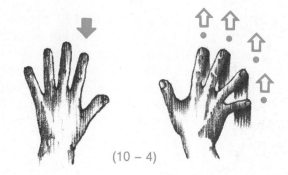

(10 − 4)

*This accumulation of 6s on top of the already established 4 is simple if you practiced the exercises outlined on page 225.

With 4 already pressed as a carryover, multiply 6 X 4. Knowing the result to be 24, add this amount.

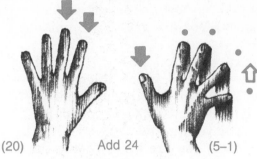

(20)　　　Add 24　　　(5–1)

You now have 28 pressed. Keep it pressed and multiply 3 X 7. Knowing the result is 21, add this amount.

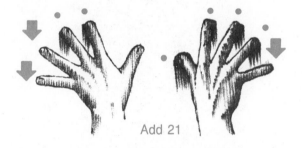

Add 21

The result, as before, is 49, but you saved many steps by knowing your facts by heart!

This simple method is not a trick. It is based upon a sound mathematical procedure that is easy to understand if one takes the time to analyze the traditional method. Using the same example:

$$
\begin{array}{r}
\overset{\text{TENS}\;\;\text{UNITS}}{}\\
4\,7\\
\times\,3\,6\\
\hline
2\,8\,2\\
1\,4\,1\\
\hline
1{,}6\,9\,2
\end{array}
$$

We must recognize that the first column of numbers represents units while the second column represents tens. (If we had a third column it would represent hundreds, and a fourth column would represent thousands.) You first multiply 6 units by 7 units. The result is 42.

Next, when you multiply 6 units by 4 tens (40) you are really getting answer of 240, which is added to the 42 already calculated, for a total of 282. When multiplying 3 tens (30) X 7 units you get 210. When multiplying 3 tens (30) X 4 tens (40) you get 1,200. Now adding 282 + 210 + 1,200, you have a grand total of 1,692. By using the Fingermath one-line method, you are simply doing the same calculations and combining partial products on your fingers.

One Liners

32 X24	82 X47	96 X12
57 X28	36 X37	82 X26
96 X14	28 X56	44 X33
62 X28	83 X29	75 X67
38 X56	83 X84	29 X63
48 X15	67 X81	32 X74
82 X28	64 X28	33 X55
97 X13	77 X38	68 X86

The same system is used for any combination of numbers. For example, if you wanted to multiply a three-digit number by a two-digit number, you would use the following sequence of moves to get a one-line answer:

$$529 \\ \times 34$$

Note in the following steps that each number is multiplied once by every other number. Units are multiplied by units, 10s and 100s; 10s are multiplied by units, 10s and 100s; and 100s are multiplied by units and 10s in the following order:

Step 1

$4 \times 9 = 36$. Enter 6, carry 3.

(36)
$$529 \\ \times 34 \\ \hline 6$$

Step 2

3 (carryover) $+ (4 \times 2) + (3 \times 9) = 38$.

Enter 8, carry 3.

(8)   (27)
$$529 \\ \times 34 \\ \hline 86$$

Step 3

3 (carryover) $+ (3 \times 2) + (4 \times 5) = 29$.

Enter 9, carry 2.

(20) (6)
$$529 \\ \times 34 \\ \hline 986$$

Step 4

2 (carryover) $+ (3 \times 5) = 17$

Enter 17.

(15)
$$529 \\ \times 34 \\ \hline 17,986$$

Always remember to keep the previous carryover pressed as you begin the multiplication of the next step.

Simple! If you think you're really sharp, try to work out the moves in multiplying other groups of numbers, using the same system of multiplying each place by every other place. Try multiplying two sets of three-digit numbers. It's a great mental exercise and the combination, once you figure it out, always works.

326 X14	837 X59	637 X28
852 X69	321 X12	918 X63
787 X32	413 X96	869 X26

Conclusion

Perhaps the most invigorating aspect of education is that it never ends.

No matter how much of it we devour, there is always more to be enjoyed. Paradoxically, the more we know, the more we understand how little we know.

It is my hope that this book of Fingermath provides just such a beginning experience, that it primed your appetite for the challenging and rewarding world of mathematics. Children who become so highly motivated and enthusiastic with this game-like involvement with arithmetic, are encouraged to seek out new levels of achievement in academic areas they may earlier had thought of as being beyond them.

I will also allow myself the privilege of believing that those of you who have found satisfaction in the methods of Fingermath have forever put behind you the notion it is wrong to count on your fingers. Finger computation is fun and is a great way to learn.

342